INTRODUCTION TO SYSTEMS BIOLOGY

INTRODUCTION TO SYSTEMS BIOLOGY

Workbook for Flipped-Classroom Teaching

Thomas Sauter and Marco Albrecht

https://www.openbookpublishers.com

ISBN Paperback: 978-1-80064-410-6
ISBN Hardback: 978-1-80064-411-3
ISBN Digital (PDF): 978-1-80064-412-0
DOI: 10.11647/OBP.0291

Cover image: Photo by Vlado Paunovic on Unsplash at https://unsplash.com/photos/LfsXOnS41dg

Cover design by Anna Gatti

Preface

The content of this book was developed over more than 15 years of teaching the course "Introduction to Systems Biology", first at the University of Stuttgart and then mainly at the University of Luxembourg. This course aims to introduce key mathematical concepts of systems biology to students with mainly biology backgrounds. Easily accessible toy examples are used to illustrate these concepts in a straightforward way. Some of these examples, as well as some of the ideas in the book, come from colleagues, whom we would like to thank very much for sharing their work.

Over the years, the course style changed from traditional classroom teaching—with lectures on the concepts and demonstrations of exercise solutions—to more self-paced and interactive learning using the flipped-classroom method (see the Introduction of this book). This usually consisted of a short kick-off lecture emphasizing the key concepts briefly and answering some general questions of the class. The remainder of the day was then organized into flexible group work in class with the support of tutors, and independent study time (usually in the afternoons). This allowed the students to progress at their own pace and to support each other. Final exam results improved by around 2 points on a scale of 20 as a result of this new method.

The course was complemented with talks about current research questions and examples of the lab or the field in general. These talks were either given by me (Thomas Sauter) or by the assisting postdoctoral and PhD students. Within the curriculum of the Master's in Integrated Systems Biology at the University of Luxembourg, this course was followed by 2 practical computational courses, where the students applied the introduced mathematical concepts to self-designed and self-executed projects. These project-based learning courses focused on metabolic network modelling using constraint-based modelling (see Chapter 2 of this book) and on pharmacokinetic (PK) modelling using ordinary differential equations (see Chapter 3 of this book). The structure of these courses, along with some illustrative example projects, is detailed in the article "Project-Based Learning Course on Metabolic Network Modelling in Computational Systems Biology" (Sauter et al., 2022)[1]. The combination of studying the theory at one's own pace and applying it to self-designed projects has proven to be an effective way of learning.

---

[1] *PLoS Comput Biol* 2022 Jan 27; 18(1):e1009711, https://doi.org/10.1371/journal.pcbi.1009711.

   https://doi.org/10.11647/OBP.0291.05

Suggestions and corrections are very welcome (by email to: thomas.sauter@uni.lu) and will be considered for the next edition of this book.

On a personal note, I would like to take this opportunity to express my thankfulness to my parents—your love and hard work have laid the foundation for my career—and to my family: Sabine, it is so precious to have you by my side. Josephine, it is great to see you growing up and shining. And Leonard, I am grateful for our days together. You were the first to see this book.

Thomas Sauter, Nittel & Belval, October 2022

Acknowledgments

The authors would like to thank several people for their contribution and support, in particular:

- Prof. Dr.-Ing. Herbert Wehlan, Institute for System Dynamics, University of Stuttgart, Germany
- Dr.-Ing. Michael Ederer
- Prof. Dr.-Ing. Andreas Kremling, Technical University of Munich, Germany
- Dr.-Ing. Steffen Klamt, Max Planck Institute for Dynamics of Complex Technical Systems, Magdeburg, Germany
- Ass.-Prof. Dr.-Ing. Steffen Waldherr, University of Vienna, Austria
- Dr. Giulia Cesi, Department of Life Sciences and Medicine, University of Luxembourg
- Dr. Maria Pires Pacheco, Department of Life Sciences and Medicine, University of Luxembourg
- Apurva Badkas, Department of Life Sciences and Medicine, University of Luxembourg
- All MISB and IMBM students who went over the course materials over the last few years.

Introduction

Thomas Sauter, Marco Albrecht

Motivation

In this book, you will learn how mathematical models of biological networks are built and how the analysis of such models help to understand the system-level properties of networks. The book will introduce you to the language of systems biology which needs to be spoken among biologists, physicists, computer scientists, and engineers in the interdisciplinary research environment of bio-medicine. Science is about what is; Engineering is about what can be. Combining both will enrich your profile as an academic and enrich your view of the world around us. We are on the brink of the era of network medicine. This novel approach has the potential to revolutionize and personalize the treatment of patients. This book focuses on some of the fundamental concepts which are essential to developing successful network medicine approaches in the upcoming years. We hope you enjoy reading this book as much as we enjoyed writing it.

Keywords

Systems biology — Flipped-classroom teaching

Contact: thomas.sauter@uni.lu. **Licence**: CC BY-NC

Contents

1. Authors

Thomas Sauter has been professor for Systems Biology and study director of the Master in Integrated Systems Biology and the International Master in Bio-Medicine at the University of Luxembourg since 2008. He studied Technical Biology at the University of Stuttgart and at the Max Planck Institute for Dynamics of Complex Technical Systems in Magdeburg, Germany. He received a PhD in Engineering for modeling of the metabolism of *Escherichia coli*. His research group develops tools for molecular network reconstruction and network-based drug discovery, with applications mainly in cancer biology. He has more than 20 years of experience in educating and supporting students.

Marco Albrecht is an engineer, trained in system theory, control engineering, modeling, and molecular biology. He studied biosystems engineering at the Otto-von-Guericke University in Magdeburg and did a PhD at the University of Luxembourg on "Mathematical histopathology and systems pharmacology of melanoma" in the context of the MELPLEX ITN training program supported by the European HORIZON 2020. He is now a research scientist at esqLABS GmbH, Germany, with expertise in Quantitative Systems Pharmacology.

2. Overview

Complex systems can be found in many fields, and researchers in biology take ever more advantage of this and related concepts shown in Figure 1. The concepts are now reaching the realm of medicine and also raise several challenges for data integration. We suggest reading the paper on systems medicine [1] which is summarized in Figure 2. Some of these concepts will be explained in this book "Introduction to Systems Biology".

Many computational courses rely on linear algebra and other mathematical concepts. Consequently, it will be very important to pay sufficient attention to these mathematical basics. We incorporated a good share of

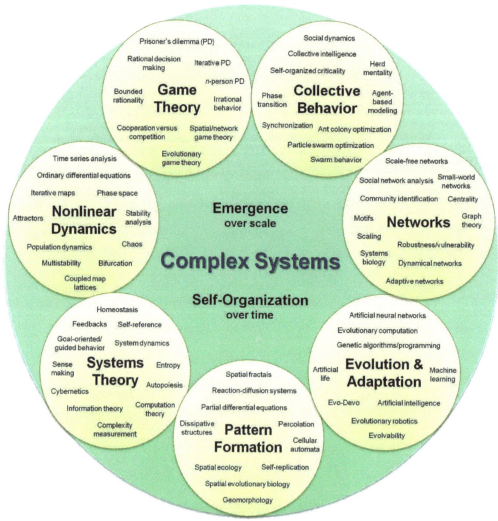

Figure 1. Complex systems organizational map. Created by H. Sayama, Collective Dynamics of Complex Systems Research Group at Binghamton University, New York. Wikimedia. Licence: CC BY 4.0.

it in this book, as you will see in the following chapters. But first we would like to make some remarks about the design of the book and the respective course on "Introduction to Systems Biology". We therefore review the research evidence for effective learning and reveal pitfalls which might emerge within an interdisciplinary study program. We also give you lists with small and prioritized learning units, which you can tick off step by step. This book contains several links to YouTube videos. Check them out by clicking on the link:
YouTube: Systems Thinking

3. Planning

We divided this course into four parts with increasing levels of modeling detail, shown in Figure 3.

The detailed content is specified in the learning checklist on page 8. Here, we give here a rough overview of what we want to achieve.

Course aims (what):

- Gain confidence in the step-wise calculation of mathematical problems.

- Connect mathematical concepts to biological real-world problems.

- Enable efficient communication between biology and computational disciplines.

Course goals (how):

- We demonstrate the step-wise calculation in this book and with the help of YouTube videos.

- We connect theoretical approaches with real-world biology.

- We explain the geometrical intuition behind mathematical operations.

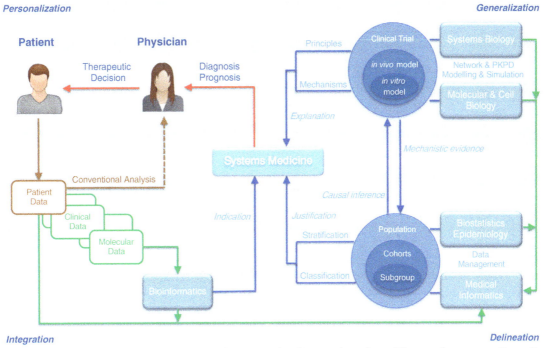

Figure 2. Systems medicine: Brown: conventional approach. Green: data flow. Blue: information flow. Source: [1]. Licence: CC BY 4.0.

This book has been developed for a full-time two-week course following the flipped-classroom approach which we will introduce in the next section.

4. Learning

The following insights come from educational studies [2]. Social scientists compare between-group differences and within-group differences with the measure *Cohens d*. A *Cohens d = 0.5* means that the difference between groups is half the difference within groups. Social scientists interpret d as follows:

$d \approx 0.2$: small effect

$d \approx 0.5$: medium effect

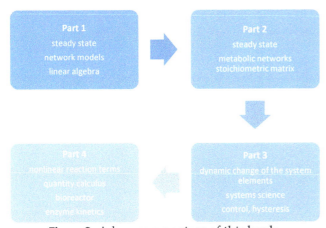

Figure 3. Advance organizer of this book.

$d \approx 0.8$: strong effect

with the hint that even minor effects can become relevant in combination with others.

Self-efficacy belief and regulation of effort

The most effective attitude is the self-efficacy belief ($d = 1.81$). Accordingly, we have organized the course in a way that ensures you have the most flexibility in tackling the problems on your own, and we will try to support you. We have tried to give clear objectives, prioritize the different tasks and optimize the course structure to help you progress fast without losing time. These precautions also complement your ability to regulate the effort by yourself ($d = 0.75$). Your positive energy and willingness to master this ambitious course will make the biggest impact beyond anything we can do. Passing this course gives you a great feeling of accomplishment and a new view of biology. With the right practice and the belief that you can make the most difference, you have the key to success in your own hands. This effect size is very strong and compensates for differences in talent, intelligence, and unchangeable traits to a large extent. Personality, intelligence, gender, time of year and working hours (for an office-based job) are altogether minor effects ($-0.24 < d < 0.32$). Intelligence explains 4% of the exam results. Joy, pride, and hope ($-0.24 < d < 0.32$) are more productive attitudes than anger, fear, and charm ($-0.8 < d < -0.28$).

Effective practice in an interdisciplinary environment

The success of teaching various learning strategies declines from elementary school ($d = 0.92$) to university ($d = 0.28$), which can be explained by the supposition that students learn which strategy is best for achieving results over time. However, learning strategies are highly subject dependent and can hardly ever be transferred to other disciplines. Studying concepts in biology requires the memorization of many facts to achieve a sufficient knowledge base. A huge amount of initially unrelated facts have to be learnt in order to interpret new observations, design experiments, and understand relationships. Mathematics and engineering, however, require the memorization of a few and simple basic concepts with which they construct their theories. Only axioms and basic equations must be learned. The challenge is to apply those concepts to different cases and tasks. Some tasks seem simple but can be unsolvable problems, while other, seemingly more complicated equation sets, can turn out to be easy. Getting a feeling for the underlying approaches in each discipline takes time. Biologists usually have to make countless observations and deconstruct things in order to understand their origin. In contrast, engineers combine different elements to build something up and to achieve a certain behavior. Engineers combine problem-dependent modules of equations together to represent desired or natural systems and their behavior. In contrast, physicists always search for a simple underlying equation to help them understand nature itself. Computer scientists, bio-informaticians etc structure, handle, and store data by automating procedures according to the wishes of a user without the inner motivation to understand nature itself. A new problem can confront computational scientists with the time-consuming need to develop new software. Once this step is solved, the computational running time for solving the actual problem might be low. Therefore, they always search for pre-developed software modules and libraries. The generation of data in biology is much more incremental and steady, partly because the problems and tasks are often unique. Thus, the general thinking and research practices of different scientists can contrast. Synthetic biology is a sub-discipline of biology which resembles the thinking in engineering the most. To engineers, it might be helpful to say that they have to solve a so-called inverse problem, which is the most frequent problem biologists face. This is a very sharp separation of different thinking schools, and you will see that scientists can have a mixture of those approaches but it might help to recognize problems of misunderstanding. Neither of these is wrong, nor better than the other. The problems they tackle have simply moulded their way of thinking to the optimal mode for the discipline, which would probably fail if applied to another area.

Do not underestimate the amount of effort required to learn mathematics. Concepts make up around 20-30% of your learning time and 70-80% of your time will be necessarily devoted to solving equations and tasks on your own. This can be best compared with your lab work. The more you can automatize isolated tasks like media preparation and pipetting, the more capacity is free to solve more comprehensive and complex working schedules in the lab. Time set aside for practice is important (see Figure 4). In the beginning, you will work through several subparts of a task, but one individual subtask might still limiting your overall performance. This can be frustrating—for example, if one learns a new programming language. At first, it seems unfathomable, but you can make more progress than you think. After you have reached a certain level, you will progress very fast. At the upper level, you will become so proficient that the improvements seem to slow down as they are not recognizable anymore. At this stage, expert feedback is necessary to help you recognize flaws and find new challenges to work on.

Additionally, having willingness to solve the given problem with different approaches, whatever it takes, is a good trait to become a good computational scientist. We provide you with the solutions directly to give you more responsibility, but do not look at the solution immediately–only if you get stuck for a long time. You have to improve your skills, not just your knowledge. One also has to frequently change between studying concepts and practicing in order to progress. Some formulations might be circuitous at first glimpse, but become more understandable after solving tasks. But don't worry, the purpose of this course is an introduction to computational problem-solving and many difficulties remain even in physics, mathematics, and engineering schools, where years are dedicated to solving such tasks. Much of what you learn

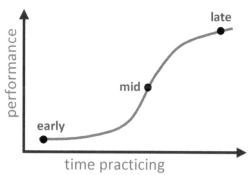

Figure 4. Performance gain in complex tasks can seem nonlinear. When too many uncertainties in subtasks hamper overall performance, sufficient practice time accelerates performance until it flattens down again. Do not give up too quickly. You never know when you will skyrocket. Source: [3]. Copyright © 2010, John Wiley and Sons.

in the course is comparable with learning a sequence of activities not far from following a cooking recipe. You will manage it! Because we integrate **active learning** sessions, we will likely reduce the failure rate. Traditional lecturing would increase the failure rate by 55% in science, engineering, and mathematics [4].

How to study engineering, math, and physics

We have some tips for studying courses with many equations. Our book will be somewhat between a classical biology and typical engineering text.

About understanding and learning

It is quite favorable to tackle the material before the lecture. Concepts in mathematics, engineering, and physics are more or less always the same and do not change as fast as some concepts in biology. They are also not as comprehensive as in biology. But they are not so easily accessible, because mathematical terms are made up of highly compressed knowledge. Lecturers in biology more often use PowerPoint presentations to transmit the knowledge, while lecturers in engineering use the blackboard to slow the knowledge transfer down. It is essential to see how things evolve. It will also be on an entirely different level than what you are used to from high school. In an engineering class, you have to plan more time for *digesting* and *understanding* the material before moving on to new topics. Most engineering students prefer to see the concepts first in order to be able to better follow the lecture content. Engineering students spend hours trying to understand the material at home. You will also need weeks and months of occasional revisiting until the material is sufficiently digested. This is the reason for the late final exam. Understanding is the biggest problem, and after you understand the material, you have to learn little by heart. Remember, you learn a lot in biology, and then you understand it. You have to understand and practice a lot in engineering, and then you learn a bit by heart.

Problem-solving

The major time-consumer will be problem-solving. You will be confronted with many tasks and problems. The more problems you solve, the better you will understand how to apply the information you have learnt and the better your grade will be. Solve the tasks we have given you! If that is not sufficient, search for more tasks in textbooks. Also solve the problems set out in past exams. It is important not to give up and to embrace the intellectual challenge. Try as many methods and strategies as possible and always look for possible calculation mistakes or typos. Messing up the minus-sign and plus-sign is quite common. Only if you get completely stuck and consultation of the theory no longer helps, then you should look up the solution.

Study groups

Everyone has times when they get stuck, and the desire to give up is strong. Establish study groups of 3-5 people to explain the issue to each other. More or fewer students than this is ineffective. Group members should have more or less the same ability level. Share insights, knowledge, and understanding of theories, formulas, and equations. Collaborative learning is beneficial, and you do not stand in competition with each other. However, do the work by yourself first to figure out how to get started. In groups, some students might be very fast, and then you do not learn how to tackle engineering problems on your own. Moreover, never end a group meeting when one member has still not understood the issue. This is a great opportunity to learn and solidify your knowledge by teaching. Find ways to achieve understanding. Maybe one has to figure out gaps in previous knowledge and then explain this. Each student should explain at the end what the problem was and how the solution has been obtained.

Be flexible and chill a bit

The general recommended sequence is:

1. Read lecture notes
2. Read books
3. Understand sample questions
4. Do the homework

Well, not many engineering students do this. Go jovially through the script and if you get stuck for more than 5 minutes, just go on. Forcing yourself to go through the script and trying to understand everything step for step has disadvantages. You might read too much, sleep away, and at the end the questions still confuse you—and time runs out. A better strategy might be to first read the questions in the exercise and try to solve them.

1. Go through lecture notes calmly
2. Look at the exercise questions and look at what you can solve already
3. Understand sample questions in the manuscript
4. Understand the manuscript explanations and search out textbooks
5. Iterate! Go back and forth
6. At the end, try to understand everything including the manuscript

If you do tasks at an earlier point in this sequence, you will get stuck for sure. One expects this without understanding the lecture notes. But now, you have a question in mind, and it will be easier for you to understand the lecture notes. After you have tried the examples, look at the sample questions, and if you struggle there, look at the lecture notes and books. Only at the end, and when you have tried everything, look at the solutions. Wait at least one day before you look up solutions. What we want to say is that you will have to use an iterative working style between example questions and theory. Do not be too strict and harsh with yourself. But of course, in the end, you should understand everything, the complete handout. Also read textbooks or related papers to get a consistent view on the issues and connect new knowl-

edge with old. This will help you later in the following courses and your career, in a way that only looking up things related to solving tasks will not.
YouTube: Nine study techniques for engineering courses
Education corner: How to study engineering?

How to study medicine and biology

Medicine and biology are characterized by a massive workload. As a former mathematics or physics student, you might not be used to this enormous amount. You might think it unnecessary to learn all this as long as you understand the underlying laws—but this is not the case. For example, immunology is so comprehensive and complex that you genuinely need to learn all this stuff before you really start to understand how the immune system works. No biologist will ever take you seriously if you do not catch up and show a decent knowledge base. Moreover, your computational models will fail if you do not know enough of all the issues and complexity around them. Even if you do not model everything, knowing the details is nevertheless crucial. Knowing more information helps you to guide your modeling better. You need excellent time management, reading skills, and memorizing strategies to manage this. You will have to read much more, and the biology books are much thicker. In these disciplines, it is also helpful to teach others. Watch the highly recommended advice of a graduate of a medical school, and the organization skills of a medical student. Their learning strategies are impressive.
YouTube: Medical School: How to study, read, and learn
YouTube: Watch an organized medical student

How to watch educational videos

Watching educational videos is not like watching a Bollywood movie.

Learning objectives: take one to two minutes to think about what learning goals you have before starting a video. Many videos, linked in this handout, help you deepen your knowledge, but do not forget to make progress. First go through the handout and then use the possibility to go deeper. Plan your learning.

Pause and ponder: if you were not concentrating for a moment or you missed the point, rewind or push the stop button.

Speed adjustment: if you can, speed up or slow down the video for your convenience. Double-speed? Why not?

Take notes: you cannot ask questions immediately. Jot your thoughts down and keep them for the lecture in the classroom. Apply the Cornell note-taking system: the upper left column (1/3) of your sheet is reserved for questions and keywords. The right

column (2/3) is used for your notes as usual. At the bottom of your sheet is a summary section (5cm). Fill the left column and the summary section in within 24 hours of taking your notes. It will help you reflect on the content.

Avoid distractions: keep distracting devices like iPods and smartphones away.

Watch in small pieces: if you watch everything at once for long periods, it is less efficient than spreading the sessions over time. Watch a video every now and then.

Enjoy with peers: you might use the opportunity to discuss the content with others so you can learn from each other.

YouTube: Cornell notes method

Self-directed learner and critical thinking

Learning habits are set out in stages, as shown by Grow's levels of self-directed learning [5].

Stage 1 (Dependent learner): Relies on instructor. No self-direction. Task-oriented.

Stage 2 (Interested learner): Not always directed. Seeks some opportunities and sets some goals.

Stage 3 (Involved learner): Ability to learn individually. Has learning goals and methods to achieve those goals.

Stage 4 (Self-directed learner): Sets goals. Knows how to assess and how to self-motivate. Finds valid and reliable resources.

YouTube: Self-directed learning (Part 1)
YouTube: Self-directed learning (Part 2)

To become a self-directed (self-regulated, lifelong) learner, you must learn to **assess** the demands of the task, **evaluate** your previous knowledge and skills, **plan** your approach, **monitor** the progress, and **adjust** the strategy if needed [3]. Planning the learning process is a step which is frequently ignored, and the time required for learning to take place is often underestimated. Ponder on why you take a certain approach and not another one. Also think about what was ineffective last time and how this can be improved in the future. Self-critical evaluation is important to avoid directing yourself meaninglessly. Keep in mind what Karl Popper[1] said: "If we are uncritical we shall always find what we want: we shall look for, and find, confirmations, and we shall look away from, and not see, whatever might be dangerous to our pet theories". Wisdom and the best approximation

---

[1] Austrian and British philosopher Sir Karl Raimund Popper 1902—1994.

of truth come only if you are your most merciless but constructive critic. It is the right but the hardest way. Also watch the lecture series on **critical thinking**, which will help you to become a better scientist.

YouTube: Critical thinking

In a book based on the work of the Foundation for Critical Thinking we found the following definition by Francis Brown[2]: "Critical thinking is a desire to seek, patience to doubt, fondness to meditate, slowness to assert, readiness to consider, carefulness to dispose and set in order; and hatred for every kind of imposture."

Repetition or elaboration strategies

Repetition does not have a significant measurable impact on learning. Repetition is the consolidation of something but this does not mean you are consolidating something useful or correct. Misconceptions can be consolidated as well. Thus, feedback from peers and the teacher is important. Much more effective is *deliberate practice* which directly targets self-identified weaknesses and requires a healthy portion of self-criticizing and critical thinking. Additionally, repeating easy tasks does not help you to become better. Search for challenges and practice annoying or difficult tasks with attainable goals. Moreover, a better strategy than repetition is elaboration. Elaboration deals with the integration of new pieces of information into your existing network of knowledge organization. Elaboration is more effective with high self-activity ($d = 0.7$) rather than letting the teacher do it for you ($d = 0.44$). Make connections to your previous knowledge instead of repeating facts alone and search for tasks which challenge you.

Approach to dealing with mistakes

Learning something new opens up space for opportunity, and if you dare to learn something new, mistakes will happen. The more mistakes you make, the more you will learn in the long term. Embracing new challenges and thus taking the risk of failure will carry you farther than avoiding challenges to avoid mistakes ($d = 0.44$). This strategy might lead to problems in the learning period ($d = -0.15$) but result in better performance after the learning period ($d = 0.56$). This approach is effective if the test is similar to the practice tasks ($d = 0.2$) and superior in applying the learned facts to new problem types ($d = 0.8$), which will help you to get even more out of this course in the future. Inaccurate prior knowledge or even misconceptions (the heart oxygenates the blood, Pluto is a planet, objects of different masses fall at different rates, blind people hear better) are difficult to repair if the teacher is unaware of them before the exam. Be considerate toward others making mistakes, and do not fear embarrassing moments yourself. Your only duty

---

[2] English philosopher, scientist, jurist, statesman, and author Francis Brown 1561—1626. Seen as father of empiricism and scientific methods.

is to learn from mistakes in order to improve your work. Careless and deliberate sloppiness has nothing to do with it and is not appreciated.

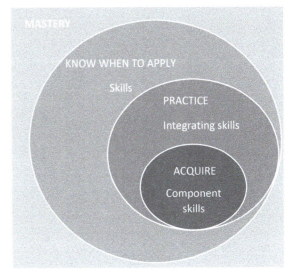

Figure 5. Elements of mastery. Source: [3]. Copyright © 2010, John Wiley and Sons.

Knowledge and skill levels

Knowledge falls into several types [3]. **Declarative knowledge** describes the knowledge of facts and concepts that can be stated or declared. **Procedural knowledge** is knowing how to apply various procedures, methods, theories, and styles. **Contextual knowledge** describes the ability to know when something has to be applied and **conceptual knowledge** says why something is appropriate in a particular situation. See also Figure 5 for the stages of mastery and Bloom's Taxonomy in the appendix [6] (Fair Use) for the classification of thinking skills.

We not only have different knowledge types, but this knowledge is also organized in different ways. The **knowledge organization** of beginners shows few connections between elements and looks like separated knowledge islands or a linear sequence of knowledge pieces, whereas experts' knowledge is densely connected—for example, in a hierarchical or network form. History facts might be memorized along a timeline, but if the question requires knowledge organized along other criteria, or the chain of knowledge is interrupted, the knowledge might be not accessible. **Mind maps** might be a good possibility of connecting pieces of knowledge in different ways. Competence can also be classified into four different stages, as shown in Figure 6. In the beginning, it is impossible to know what one has never learned before. After a while, one recognizes knowledge gaps and fills them until the acquisition process and origin get lost. Professors are frequently in the top competence level and may find it difficult to identify the problems with which you struggle. Participating during lectures and explaining questions clearly can help your supervisor to help you become better.

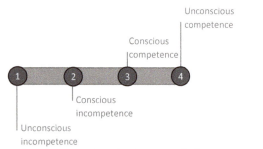

Figure 6. Competence levels. Source: [3]. Copyright © 2010, John Wiley and Sons.

Summary
Effect of educational strategies [2]
$d = 1.81$ Self-efficacy belief
$d = 1.39$ Preparation and planning by the lecturer
$d = 1.35$ Clear and understandable lecturer
$d = 1.13$ Deliberate practice objectives
$d = 1.12$ Clear learning objectives
$d = 0.90 - 0.98$ Attending courses regularly
$d = 0.77$ Openness to outsider opinions
$d = 0.75$ Student's regulation of effort
$d = 0.68$ Group work
$d = 0.68$ Empathy and warm-heartedness of the teacher
$d = 0.67$ Previous skill level
$d = 0.65$ Teacher-student relationship quality
$d = 0.64$ Co-operative learning
$d = 0.57$ Teacher's enthusiasm
$d = 0.49$ Diligence
$d = 0.48$ Motivation
$d = 0.47$ Supportive atmosphere in the classroom
$d = 0.47$ Encouragement of learning
$d = 0.43$ Solely making notes during oral presentations
$d = 0.41$ Addressing learning progress
$d = 0.41$ Organized learning
$d = 0.41$ Learning with fellow students
$d = 0.41$ Time management
$d = 0.34$ Disturbances during lessons
$d = 0.32$ Critical thinking
$d = 0.32$ Intrinsic motivation
$d = 0.21$ Class size
$d = 0.21$ Compulsory attendance
$d = 0.19$ Co-teaching
$d = -0.02$ Making notes during PowerPoint presentations
$d = -0.43$ Fear exams
Disturbing effects on exam results variability is minor (n=403623 students, 911 effects investigated):
6 % Procrastination
5 % Diligence
4 % Fear of exams
4 % Intelligence
3 % Emotional intelligence
2 % Socio-economic status
<2 % Biological age
<2 % Gender
<2 % Extroversion
<2 % Self-esteem
<2 % Social support
<2 % Stress
<2 % Depression

5. Learning checklist

Check boxes if appropriate. It might help you not to forget things and might inspire you to promote your self-directed learning. Try to stay within the script and do not lose too much time by finding answers. Use it as an inspiration and organization tool. You should ask yourself the following questions [7]:
- What is it and how is it defined? (declarative)
- How is this theory applied in the real world? (procedural)
- Could you provide an example of when this formula might be used? (contextual)
- Could you sketch what that (solution, device, etc.) might look like? (procedural)
- How is this equation applied in practice? (procedural)
- Where did that formula come from? (conceptual)
- Do you understand when that formula is used? (contextual)

Remember [3]: **Declarative knowledge** describes the knowledge of facts and concepts that can be stated or declared. **Procedural knowledge** is knowing how to apply various procedures, methods, theories, and styles. **Contextual knowledge** describes the ability to know when and in which context something has to be applied and **conceptual knowledge** says why something is appropriate in a particular situation. Does the concept fit your application? Do you know the concept behind a certain definition?

Another definition can be found (trainingindustry.com):

Definition 1. Conceptual knowledge refers to the knowledge of, or understanding of concepts, principles, theories, models, classifications, etc. We learn conceptual knowledge through reading, viewing, listening, experiencing, or thoughtful, reflective mental activity.

Definition 2. Declarative knowledge refers to facts or information stored in the memory, that is considered static in nature. Declarative knowledge, also referred to as conceptual, propositional or descriptive knowledge, describes things, events, or processes, their attributes, and their relation to each other. It is contrary to procedural, or implicit knowledge, which refers to the knowledge of how to perform or operate.

Definition 3. Procedural knowledge refers to the knowledge of how to perform a specific skill or task, and is considered knowledge related to methods, procedures, or operation of equipment. Procedural knowledge is also referred to as implicit knowledge, or know-how.

Definition 4. Implicit knowledge is knowledge that is gained through incidental activities, or without awareness that learning is occurring. Some examples of implicit knowledge are knowing how to walk, run, ride a bicycle, or swim.

Example:
The determinant of a 2-by-2 matrix is the area between two linear independent vectors (declarative). It can be computed in the following ways (procedural). The determinant is useful to understand whether a matrix is invertible (contextual) and only works if the matrix is a square matrix (contextual). The determinant is based on the geometric intuitions and concepts of linear algebra in the following way (conceptual).

Part 0: Introduction and learning

priority	tasks	first read	second read	watched YouTube	solved tasks	solved extra tasks	declarative	procedural	contextual	conceptual
			steps				knowledge			
2	Systems medicine	✓								
2	Network medicine									
2	Self-efficacy belief									
2	Regulation of effort									
2	Performance gain-practice									
2	Self-directed learner									
2	Critical thinking									
2	Knowledge organization									
2	Elaboration strategy									
2	Skill level									
2	Elements of mastery									
2	Competence level									

Part 1: Biochemical network in the matrix form

priority	tasks	first read	second read	watched YouTube	solved tasks	solved extra tasks	declarative	procedural	contextual	conceptual
			steps				knowledge			
1	Define systems biology	✓								
2	Incidence matrix									
2	Adjacency matrix & list									
3	Graph notation (brackets)									
2	Hypergraph									
1	PCA, PLSR, VIP									
1	Turn linear equation set to matrix form									
1	Matrix indices									
1	Augmented coefficient matrix									
2	Solve equations: Rule of Cramer									
1	Gauss and Gauss Jordan form									
3	Reduced row-echelon form									
3	LU decomposition									
1	Rank									
1	Identity matrix									
1	Zero matrix									
1	Trace									
1	Matrix multiplication									
1	Sum and subtract matrices									
1	Scalar multiplication									
1	Transpose									
1	Determinant of a 2-by-2 matrix									
2	Determinant of a 3-by-3 matrix (Rule of Sarrus)									
2	Determinant (Rule of Cramer)									
3	Laplace expansion									
1	Inversion of a 2-by-2 matrix									
2	Inversion of a 3-by-3 matrix									
1	Eigenvalues									
2	Eigenvectors									
3	Eigenvalue via fast equation									

Part 2: Metabolic modeling

priority	tasks	first read	second read	watched YouTube	solved tasks	solved extra tasks	declarative	procedural	contextual	conceptional
1	Classify metabolic models	✓								
1	Stoichiometric matrix									
1	Steady state									
1	Rouché–Capelli theorem									
2	Elementary flux modes (EFM)									
2	Conservation relations									
3	Left and right null space									
1	Classify MFA according dynamic and isotope tracer									
2	Metabolic flux analysis (MFA)									
1	Pros and cons of FBA									
2	Flux balance analysis (FBA)									
2	Constrained optimization cone									

Part 3: The magic of change and how to find it

priority	tasks	first read	second read	watched YouTube	solved tasks	solved extra tasks	declarative	procedural	contextual	conceptional
1	Black box concept	✓								
2	Hysteresis									
1	Block diagram									
2	Synthetic Biology vs. Systems Biology									
1	ODE									
2	What is the difference between ODE & PDE									
2	Change one ODE to a system of ODEs									
1	General properties of a system									
2	Nonlinear dynamic									
2	Open loop vs closed loop									
1	Feed-forward loops									

priority	tasks	first read	second read	watched YouTube	solved tasks	solved extra tasks	declarative	procedural	contextual	conceptional
1	Feedback loops									
1	State space representation									
1	Classify system types									
2	SISO vs MIMO									
3	Laplace transform and frequency domain									
3	Fourier transform									
2	Time domain vs frequency domain									
2	Controllability									
2	Observability									
2	Transfer function									
1	Definition steady state									
1	Stability									
1	Damping									
1	Characteristic polynomial									
1	Eigenvalues in the frequency domain to stability classification									
2	Phase portrait									
2	Definition trajectory									
2	Slope field									
2	Definition isoclines									
2	Discrete in state and time									
2	Difference equation									
1	p-q equation									
1	a-b-c equation									
1	Complex numbers									
1	Differentiation									
1	Product rule									
1	Quotient rule									
1	Chain rule									
1	Separation of variables									
1	Integration factor									
2	Linearization									

Part 4: Physical modeling and nonlinear enzyme kinetics

priority	tasks	first read	second read	watched Youtube	solved tasks	solved extra tasks	declarative	procedural	contextual	conceptional
1	Reality and model	✓								
1	Modeling cycle of Blum and Leiß									
1	Assumptions									
1	Model building									
f1	Distinguish variable, parameter, coefficient									
3	Dimension analysis									
3	Poorly posed problems									
1	Sensitivity analysis									
1	Model classification									
2	Akaike information criterion									
2	1st law of thermodynamics									
2	2nd law of thermodynamics									
3	Noise									
1	Extensive quantities									
1	Intensive quantities									
2	Read SI units									
1	Quantity calculus									
1	Balancing									
1	Mass balance									
3	Volume balance									
1	Amount of substance balance									
1	Law of mass-action									
1	Reaction rate									
1	Michaelis-Menten									
3	Lineweaver-Burk plot									
2	MM for reversible reactions									
2	MM for inhibition									
2	Substrate inhibition									
2	Cooperative enzymes									
1	Hill kinetic									
1	Mathematical analysis approaches (Wolkenhauer) without equations									

6. Further reading

We suggest reading the great article by *Barabasi et al.* to familiarize yourself with the topic of network medicine [8].

References

[1] Rolf Apweiler, Tim Beissbarth, Michael R Berthold, Nils Blüthgen, Yvonne Burmeister, Olaf Dammann, Andreas Deutsch, Friedrich Feuerhake, Andre Franke, Jan Hasenauer, et al. Whither systems medicine? *Experimental & Molecular Medicine*, 50(3):e453, 2018. https://doi.org/10.1038/emm.2017.290.

[2] Michael Schneider and Maida Mustafić. *Gute Hochschullehre: Eine evidenzbasierte Orientierungshilfe: Wie man Vorlesungen, Seminare und Projekte effektiv gestaltet.* Springer-Verlag, 2015. https://doi.org/10.1007/978-3-662-45062-8.

[3] Susan A Ambrose, Michael W Bridges, Michele DiPietro, Marsha C Lovett, and Marie K Norman. *How learning works: Seven research-based principles for smart teaching.* John Wiley & Sons, 2010.

[4] Scott Freeman, Sarah L Eddy, Miles McDonough, Michelle K Smith, Nnadozie Okoroafor, Hannah Jordt, and Mary Pat Wenderoth. Active learning increases student performance in science, engineering, and mathematics. *Proceedings of the National Academy of Sciences*, 111(23):8410–8415, 2014. https://doi.org/10.1073/pnas.1319030111.

[5] Gerald O Grow. Teaching learners to be self-directed. *Adult Education Quarterly*, 41(3):125–149, 1991.

[6] Center Grove, June 2018. https://www.centergrove.k12.in.us/Page/7844.

[7] Education Corner, July 2018. https://www.educationcorner.com/engineering-study-skills-guide.html.

[8] Albert-László Barabási, Natali Gulbahce, and Joseph Loscalzo. Network medicine: a network-based approach to human disease. *Nature Reviews Genetics*, 12(1):56, 2011. https://doi.org/10.1038/nrg2918.

Knowledge

Recall /regurgitate facts without understanding. Exhibits previously learned material by recalling facts, terms, basic concepts and answers.

Key words:

Choose	Observe
Copy	Omit
Define	Quote
Duplicate	Read
Find	Recall
How	Recite
Identify	Recognise
Label	Record
List	Relate
Listen	Remember
Locate	Repeat
Match	Reproduce
Memorise	Retell
Name	Select
Show	
Spell	
State	
Tell	
Trace	
What	
When	
Where	
Which	
Who	
Why	
Write	

Actions:

Describing
Finding
Identifying
Listing
Locating
Naming
Recognising
Retrieving

Outcomes:

Definition
Fact
Label
List
Quiz
Reproduction
Test
Workbook
Worksheet

Questions:

Can you list three ...?
Can you recall ...?
Can you select ...?
How did _____ happen?
How is ...?
How would you describe ...?
How would you explain ...?
How would you show ...?
What is ...?
What did _____ happen?
When did ...?
When did _____ happen?
Where is ...?
Which one ...?
Who was ...?
Who were the main ... ?
Why did ...?

Comprehension

To show understanding finding information from the text. Demonstrating basic understanding of facts and ideas.

Key words:

Ask	Extend
Cite	Generalise
Classify	Give examples
Compare	Illustrate
Contrast	illustrate
Demonstrate	Indicate
Discuss	Infer
Estimate	Interpret
Explain	Match
Express	Observe
Outline	
Predict	
Purpose	
Relate	
Rephrase	
Report	
Restate	
Review	
Show	
Summarise	
Translate	

Actions:

Classifying
Comparing
Exemplifying
Explaining
Inferring
Interpreting
Paraphrasing
Summarising

Outcomes:

Collection
Examples
Explanation
Label
List
Outline
Quiz
Show and tell
Summary

Questions:

Can you explain what is happening . . . what is meant . . .?
How would you classify the type of ...?
How would you compare ...?contrast ...?
How would you rephrase the meaning ...?
How would you summarise ...?
What can you say about ...?
What facts or ideas show ...?
What is the main idea of ...?
Which is the best answer ...?
Which statements support ...?
Will you state or interpret in your own words ...?

Application

To use in a new situation. Solving problems by applying acquired knowledge, facts, techniques and rules in a different way.

Key words:

Act	Employ
Administer	Experiment with
Apply	Group
Associate	Identify
Build	Illustrate
Calculate	Interpret
Categorise	Interview
Choose	Link
Classify	Make use of
Connect	Manipulate
Construct	Model
Correlation	Organise
Demonstrate	Perform
Develop	Plan
Dramatise	
Practice	
Relate	
Represent	
Select	
Show	
Simulate	
Solve	
Summarise	
Teach	
Transfer	
Translate	
Use	

Actions:

Carrying out
Executing
Implementing
Using

Outcomes:

Demonstration
Diary
Illustrations
Interview
Journal
Performance
Presentation
Sculpture
Simulation

Questions:

How would you use ...?
What examples can you find to ...?
How would you solve _____ using what you have learned ...?
How would you organise _____ to show ...?
How would you show your understanding of ...?
What approach would you use to ...?
How would you apply what you learned to develop ...?
What other way would you plan to ...?
What would result if ...?
Can you make use of the facts to ...?
What elements would you choose to change ...?
What facts would you select to show ...?
What questions would you ask in an interview with ...?

Analysis

To examine in detail. Examining and breaking information into parts by identifying motives or causes; making inferences and finding evidence to support generalisations.

Key words:

Analyse	Examine
Appraise	Find
Arrange	Focus
Assumption	Function
Breakdown	Group
Categorise	Highlight
Cause and effect	In-depth discussion
Choose	Inference
Classify	Inspect
Differences	Investigate
Discover	Isolate
Discriminate	List
Dissect	Motive
Distinction	Omit
Distinguish	Order
Divide	Organise
Establish	Point out
Prioritize	
Question	
Rank	
Reason	
Relationships	
Reorganise	
Research	
See	
Select	
Separate	
Similar to	
Simplify	
Survey	
Take part in	
Test for	
Theme	
Comparing	

Actions:

Attributing
Deconstructing
Integrating
Organising
Outlining
Structuring

Outcomes:

Abstract
Chart
Checklist
Database
Graph
Mobile
Report
Spread sheet
Survey

Questions:

What are the parts or features of ...?
How is _____ related to ...?
Why do you think ...?
What is the theme ...?
What motive is there ...?
Can you list the parts ...?
What inference can you make ...?
What conclusions can you draw ...?
How would you classify ...?
How would you categorise ...?
Can you identify the difference parts ...?
What evidence can you find ...?
What is the relationship between ...?
Can you make a distinction between ...?
What is the function of ...?
What ideas justify ...?

Synthesis

To change or create into something new. Compiling information together in a different way by combining elements in a new pattern or proposing alternative solutions.

Key words:

Adapt	Estimate
Add to	Experiment
Build	Extend
Change	Formulate
Choose	Happen
Combine	Hypothesise
Compile	Imagine
Compose	Improve
Construct	Innovate
Convert	Integrate
Create	Invent
Delete	Make up
Design	Maximise
Develop	Minimise
Devise	Model
Discover	Modify
Discuss	Original
Elaborate	Originate
Plan	
Predict	
Produce	
Propose	
Reframe	
Revise	
Rewrite	
Simplify	
Solve	
Speculate	
Substitute	
Suppose	
Tabulate	
Test	
Theorise	
Think	
Transform	
Visualise	

Actions:

Constructing
Designing
Devising
Inventing
Making
Planning
Producing

Outcomes:

Advertisement
Film
Media product
New game
Painting
Plan
Project
Song
Story

Questions:

What changes would you make to solve ...?
How would you improve ...?
What would happen if ...?
Can you elaborate on the reason ...?
Can you propose an alternative ...?
Can you invent ...?
How would you adapt _____ to create a different ...?
How could you change (modify) the plot (plan) ...?
What could be done to minimise (maximise) ...?
What way would you design ...?
Suppose you could _____ what would you do ...?
How would you test ...?
Can you formulate a theory for ...?
Can you predict the outcome if ...?
How would you estimate the results for ...?
What facts can you compile ...?
Can you construct a model that would change ...?
Can you think of an original way for the ...?

Evaluation

To justify. Presenting and defending opinions by making judgements about information, validity of ideas or quality of work based on a set of criteria.

Key words:

Agree	Disprove
Appraise	Dispute
Argue	Effective
Assess	Estimate
Award	Evaluate
Bad	Explain
Choose	Give reasons
Compare	Good
Conclude	Grade
Consider	How do we know?
Convince	Importance
Criteria	Infer
Criticise	Influence
Debate	Interpret
Decide	Judge
Deduct	Justify
Defend	Mark
Determine	
Measure	
Opinion	
Perceive	
Persuade	
Prioritise	
Prove	
Rate	
Recommend	
Rule on	
Select	
Support	
Test	
Useful	
Validate	
Value	
Why	

Actions:

Attributing
Checking
Deconstructing
Integrating
Organising
Outlining
Structuring

Outcomes:

Abstract
Chart
Checklist
Database
Graph
Mobile
Report
Spread sheet
Survey

Questions:

Do you agree with the actions/outcomes ...?
What is your opinion of ...?
How would you prove/disprove ...?
Can you assess the value/importance of ...?
Would it be better if ...?
Why did they (the character) choose ...?
What would you recommend ...?
How would you rate the ...?
What would you cite to defend the actions ...?
How would you evaluate ...?
How could you determine ...?
What choice would you have made ...?
What would you select ...?
How would you prioritise ...?
What judgement would you make about ...?
Based on what you know, how would you explain ...?
What information would you use to support the view ...?
How would you justify ...?
What data was used to make the conclusion ...?

Bloom's Taxonomy: Teacher Planning Kit

Notes

Notes

Chapter 1: Biochemical networks in the matrix form

Thomas Sauter, Marco Albrecht

Motivation

The biochemistry of the cell is very complex and the available data might overwhelm the abilities of interpretation [1]. Reductionist approaches, combined with some intuition, have brought us far, but we need rational approaches to better understand the interplay of molecules at the system level. We have to check whether a hypothesis is in itself logical and can be aligned with data. In this chapter, we—while reducing biochemical molecules to their function—will learn how to interlink several players to acquire a mechanistic understanding of a pathway or a complex system. Modeling thereby helps us in the following ways [2]:

1. Enhancing understanding of otherwise unintelligible systems
2. Requiring a way of thinking that can be beneficial to the design of experiments

While, by studying this chapter, you will not become a computational scientist, it will help you to communicate with them. Nobody expects that you understand everything immediately. It will take time to digest and it requires a lot of practicing to build the skills.

The mathematical principles introduced here will be applied to biological pathways and networks in the following chapters. If you prefer, you could directly jump to Chapter 2 and 3 to see some applications first.

Keywords

Matrix — Graph — Metabolic network

Contact: thomas.sauter@uni.lu. **Licence**: CC BY-NC

Contents

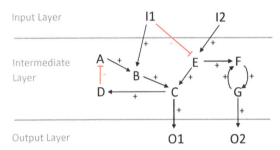

Figure 1. The interaction graph of a signal transduction pathway triggered by the inputs I1 and I2 with subsequent response in the output layer indicated by O1 and O2.

1. Lecture summary

1.1 ■ Biological networks and graph theory

Biological phenomena are very complex and systems biology helps us to understand their system's behavior.

Definition 1. Systems biology is the science that studies how biological function emerges from the interactions between the components of living systems and how these emergent properties enable and constrain the behaviour of the components [2].

Put in another way: we are not interested in dissecting objects into ever smaller parts and details. Instead, we look at the elements (nodes, states) we know and focus on their interactions (edges, coefficients). The interplay of a few elements can result in manifold different phenomena and observations, depending on how the elements activate or inhibit each other. The interactions between the states in a system are often represented as a matrix. But the type of matrix used can vary for different systems.

https://doi.org/10.11647/OBP.0291.01

Example 1: Signal transduction

The example graph in Figure 1 can be written in the form of an incidence matrix In

$$In = \begin{bmatrix} -1 & -1 & 0 & 0 & 0 & 0 & 0 & 0 & 0 & 0 & 0 & 0 & 0 \\ 0 & 0 & -1 & 0 & 0 & 0 & 0 & 0 & 0 & 0 & 0 & 0 & 0 \\ 0 & 0 & 0 & -1 & 0 & 0 & 1 & 0 & 0 & 0 & 0 & 0 & 0 \\ 1 & 0 & 0 & 1 & -1 & 0 & 0 & 0 & 0 & 0 & 0 & 0 & 0 \\ 0 & 0 & 0 & 0 & 1 & -1 & 0 & 1 & 0 & 0 & 0 & -1 & 0 \\ 0 & 0 & 0 & 0 & 0 & 1 & -1 & 0 & 0 & 0 & 0 & 0 & 0 \\ 0 & 1 & 1 & 0 & 0 & 0 & 0 & -1 & -1 & 0 & 0 & 0 & 0 \\ 0 & 0 & 0 & 0 & 0 & 0 & 0 & 1 & -1 & 1 & 0 & 0 & 0 \\ 0 & 0 & 0 & 0 & 0 & 0 & 0 & 0 & 1 & -1 & 0 & -1 \\ 0 & 0 & 0 & 0 & 0 & 0 & 0 & 0 & 0 & 0 & 1 & 0 \\ 0 & 0 & 0 & 0 & 0 & 0 & 0 & 0 & 0 & 0 & 0 & 1 \end{bmatrix} \begin{matrix} I1 \\ I2 \\ A \\ B \\ C \\ D \\ E \\ F \\ G \\ O1 \\ O2 \end{matrix}$$

with interactions as column entries and states as row entries.

Example 1 mimics a signal transduction network within a cell (Figure 1), where relevant molecules are represented as nodes (states) and interactions as edges (later resulting in mathematical terms in the balance equations). The states represent the phosphorylation status or the concentration of a particular molecule, while the interactions represent binding affinities, regulatory interactions, or metabolic fluxes etc. If molecule A is directly responsible for a higher activity or abundance of molecule B, we draw an arrow from A to B, which is called a directed edge. Moreover, we write in the related column of the incidence matrix In (see Example 1) the number -1 for A and $+1$ for B. Molecule B, on the other hand, has a positive impact on C. Molecule A would thus indirectly lead to higher levels of activation of C, but no direct interaction, so this is not represented in the network. Edges can also represent inhibitory interactions, which are drawn as a straight line with a transverse line at the inhibited molecule. The true interactions can be figured out through experimental studies or via the analysis of the overall behavior of a network. Biological systems can also be represented in the form of an adjacency matrix A or as an adjacency list L, tackled in Case Box 1 and 2. The combination of elements and interactions makes up a graph or network. Protein interactions can be represented as undirected networks. One valuable source for such networks is, for example, the STRING database [3].

We want to compare a directed network with an undirected network by reference to the cases in Figure 2. We note down the related matrices in Case Box 1 and 2.

Graphs are a special case of more general hypergraphs, shown in Figure 3. In a graph, edges connect 2 nodes, whereas in a hypergraph $\mathbf{H} = (\mathbf{V}, \mathbf{E})$ there is a set of hyperedges $\mathbf{E}$ connecting a set of vertices $\mathbf{V}$. In other words, in a hypergraph, a hyperedge can connect any number of vertices. Hypergraphs are used, for example, used to represent metabolic networks where reactions can connect multiple substrates and products and sometimes involve cofactors. Undirected hypergraphs represent set systems, as shown in Figure 4. Directed hypergraphs have hyperedges $\mathbf{e} = (\mathbf{S}, \mathbf{K})$ with vertices assigned to the tail/start knot $\mathbf{S}$ and vertices assigned to the head/end knot $\mathbf{K}$. An example directed graph is shown in Figure 5. One of the hyperedges points from the tail knots A and B ($\mathbf{S} = \{A, B\}$) to the head knots C and D ($\mathbf{K} = \{C, D\}$) written as $e_1 = (\{A, B\}, \{C, D\})$. Pay attention to the brackets.

Case 1: Undirected graph

An undirected graph is described by vertices $\mathbf{V} = \{A, B, C, D, E, F\}$ and edges $\mathbf{E} = \{a, b, c, d, e, f, g\} = \{(A, B), (B, C), (C, D), (D, E), (D, F), (C, F), (F, A)\}$. The relevant matrices (see text) are:

$$A = \begin{matrix} & A & B & C & D & E & F \\ A & 0 & 1 & 0 & 0 & 0 & 1 \\ B & 1 & 0 & 1 & 0 & 0 & 0 \\ C & 0 & 1 & 0 & 1 & 0 & 1 \\ D & 0 & 0 & 1 & 0 & 1 & 1 \\ E & 0 & 0 & 0 & 1 & 0 & 0 \\ F & 1 & 0 & 1 & 1 & 0 & 0 \end{matrix}, \quad L = \begin{matrix} B,F & A \\ A,C & B \\ B,D,F & C \\ C,E,F & D \\ D & E \\ A,C,D & F \end{matrix}$$

$$In = \begin{matrix} & a & b & c & d & e & f & g \\ A & 1 & 0 & 0 & 0 & 0 & 0 & 1 \\ B & 1 & 1 & 0 & 0 & 0 & 0 & 0 \\ C & 0 & 1 & 1 & 0 & 0 & 1 & 0 \\ D & 0 & 0 & 1 & 1 & 1 & 0 & 0 \\ E & 0 & 0 & 0 & 1 & 0 & 0 & 0 \\ F & 0 & 0 & 0 & 0 & 1 & 1 & 1 \end{matrix}$$

whereby the adjacency matrix A is symmetric. It is not symmetric for a directed graph.

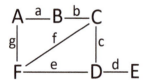 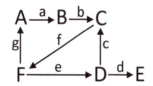

Figure 2. Left: undirected graph. **Right:** directed graph.

Case 2: Directed graph

The directed graph is represented by vertices $\mathbf{V} = \{A, B, C, D, E, F\}$ and by edges $\mathbf{E} = \{a, b, c, d, e, f, g\}$ $= \{(A,B), (B,C), (D,C), (D,E), (F,D), (C,F), (F,A)\}$. The order of vertices in the edge description is relevant now. The matrices are:

$$\mathbf{A} = \begin{array}{c} \\ \end{array}\begin{matrix} A & B & C & D & E & F \\ \end{matrix}\\ \begin{bmatrix} 0 & 1 & 0 & 0 & 0 & 0 \\ 0 & 0 & 1 & 0 & 0 & 0 \\ 0 & 0 & 0 & 0 & 0 & 1 \\ 0 & 0 & 1 & 0 & 1 & 0 \\ 0 & 0 & 0 & 0 & 0 & 0 \\ 1 & 0 & 0 & 1 & 0 & 0 \end{bmatrix}\begin{matrix} A \\ B \\ C \\ D \\ E \\ F \end{matrix}, \quad \mathbf{L} = \begin{bmatrix} B \\ C \\ F \\ C,E \\ - \\ A,D \end{bmatrix}\begin{matrix} A \\ B \\ C \\ D \\ E \\ F \end{matrix},$$

$$\mathbf{In} = \begin{array}{c} \\ \end{array}\begin{matrix} a & b & c & d & e & f & g \\ \end{matrix}\\ \begin{bmatrix} -1 & 0 & 0 & 0 & 0 & 0 & 1 \\ 1 & -1 & 0 & 0 & 0 & 0 & 0 \\ 0 & 1 & 1 & 0 & 0 & -1 & 0 \\ 0 & 0 & -1 & -1 & 1 & 0 & 0 \\ 0 & 0 & 0 & 1 & 0 & 0 & 0 \\ 0 & 0 & 0 & 0 & -1 & 1 & -1 \end{bmatrix}\begin{matrix} A \\ B \\ C \\ D \\ E \\ F \end{matrix}$$

hypergraph   substrate graph   bipartite graph

Figure 3. A hypergraph can be translated into a substrate graph or a bipartite graph. A substrate graph cannot be converted back to the hypergraph because the information whether A AND B are consumed by the same reaction is lost in the substrate graph. It could also be possible that A is converted into C and D, and that B is transformed into C and D by independent reactions.

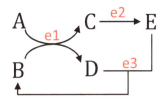

Figure 5. Directed hypergraph $\mathbf{V} = \{A,B,C,D,E\}$, $\mathbf{E} = \{e1, e2, e3\} = \{(\{A,B\},\{C,D\}), (\{C\},\{E\}), (\{E,D\},\{B\})\}$.

Example 2: Simple hypergraph

The example in Figure 5 can be represented by the incidence matrix:

$$\mathbf{In} = \begin{array}{c} \\ \end{array}\begin{matrix} e1 & e2 & e3 \\ \end{matrix}\\ \begin{bmatrix} -1 & 0 & 0 \\ -1 & 0 & 1 \\ 1 & -1 & 0 \\ 1 & 0 & -1 \\ 0 & 1 & -1 \end{bmatrix}\begin{matrix} A \\ B \\ C \\ D \\ E \end{matrix}$$

Another representation of the system in Figure 5 can be realized with chemical reaction equations:

$$A + B \xrightarrow{e1} C + D$$
$$C \xrightarrow{e2} E$$
$$D + E \xrightarrow{e3} B$$

Directed hypergraphs, as mentioned previously, are needed for an important field in systems biology: metabolic network modeling. The only difference is that we have additional stoichiometric information (coefficients) to weight the edges.

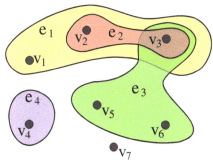

Figure 4. Undirected hypergraph: $\mathbf{V} = \{v_1, v_2, v_3, v_4, v_5, v_6, v_7\}$ and $\mathbf{E} = \{e_1, e_2, e_3, e_4\} = \{\{v_1, v_2, v_3\}, \{v_2, v_3\}, \{v_3, v_5, v_6\}, \{v_4\}\}$. Source: en.wikipedia.org/wiki/Hypergraph, Fair Use.

1.2 ■ Modeling of metabolic networks

Metabolic networks describe the flux of metabolites within a system. The fluxes are controlled by enzymes. A simplified metabolic network is shown in Figure 6 and Example 3. A more elaborate example is shown in Figure 7 and Example 4.

Example 3: Simple metabolic network

We see in Figure 6, that we have two possible pathways here, either via v_2 or v_3. All edges can be weighted according to the stoichiometric coefficients so that we do not simply have an incidence matrix (all the entries are 1s or 0s) but a stoichiometric matrix:

$$N = \begin{bmatrix} v_1 & v_2 & v_3 & v_4 \\ 1 & -1 & -1 & 0 \\ 0 & 1 & 1 & -1 \\ -1 & 0 & 0 & 0 \\ 0 & 0 & 0 & 1 \end{bmatrix} \begin{matrix} A_{in} \\ B_{in} \\ A_{out} \\ B_{out} \end{matrix} .$$

where the reactions determine the columns and the metabolite concentration determines the row entries. Often we focus on intracellular metabolites only, so that we can reduce the system in this case to:

$$N = \begin{bmatrix} v_1 & v_2 & v_3 & v_4 \\ 1 & -1 & -1 & 0 \\ 0 & 1 & 1 & -1 \end{bmatrix} \begin{matrix} A \\ B \end{matrix}$$

This example has only unimolecular reactions.

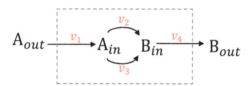

Figure 6. A simple model of a metabolic system. The gray dotted line represents the system boundary. A molecule comes from **out**side to **in**side and turns into internal Molecule B via two possible reaction ways. Molecule B is leaving the system.

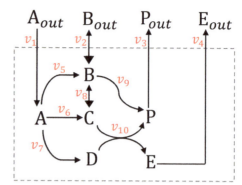

Figure 7. A metabolic model with 2 inputs and 3 outputs, whereby Molecule B is both. The system is controlled by the transport reactions of the external Metabolites A and B to the cell, which releases Product P but can also release B and E.

Example 4: Larger metabolic network

Here, we see several possible routes (Figure 7) and also a hyperedge v_{10}. It is quite difficult without mathematics to understand and predict the fluxes, which we will learn in the next learning block. For now we want to set up the stoichiometric matrix:

$$N = \begin{bmatrix} v_1 & v_2 & v_3 & v_4 & v_5 & v_6 & v_7 & v_8 & v_9 & v_{10} \\ 1 & 0 & 0 & 0 & -1 & -1 & -1 & 0 & 0 & 0 \\ 0 & 1 & 0 & 0 & 1 & 0 & 0 & -1 & -1 & 0 \\ 0 & 0 & 0 & 0 & 0 & 1 & 0 & 1 & 0 & -1 \\ 0 & 0 & 0 & 0 & 0 & 0 & 1 & 0 & 0 & -1 \\ 0 & 0 & 0 & -1 & 0 & 0 & 0 & 0 & 0 & 1 \\ 0 & 0 & -1 & 0 & 0 & 0 & 0 & 0 & 1 & 1 \end{bmatrix} \begin{matrix} A \\ B \\ C \\ D \\ E \\ P \end{matrix}$$

with the reversible reactions:

$$\text{rev} = \{R2, R8\}$$

and irreversible reactions:

$$\text{irrev} = \{R1, R3, R4, R5, R6, R7, R9, R10\}$$

After seeing some motivating examples of biological networks and the possibility of representing these as matrices, we will now revise some basic mathematical concepts of linear algebra and matrix calculation.

2. Basics of Mathematics

2.1 ▦ Linear algebra

Linear algebra (Arabic: *al-jabr*) is one of the most fundamental and helpful topics from the realm of mathematics. We recommend watching the following YouTube videos and more from that channel the channel the videos belong to, by clicking on the link below:

YouTube: The essence of linear algebra
YouTube: How to read math?

▦ From a set of equations to a matrix

In science and technology, we frequently encounter sets of linear equations. One equation might be $3x_1 + x_2 = -2$, which we can write in a general form with $a_{11}x_1 + a_{12}x_2 = b_1$ with **coefficient** $a_{11} = 3$, $a_{12} = 1$ and constant $b_1 = -2$. The **variables**, x, can represent molecule concentrations, and coefficients can represent interactions between molecules. If we add at least one other linear equation—such as $2x_1 + 1x_2 = 0$—with at least one common variable, we have a **linear equation set**:

$$a_{11}x_1 + a_{12}x_2 = b_1$$
$$a_{21}x_1 + a_{22}x_2 = b_2$$

You see that the first **index** m of coefficient a_{mn} increases with the number in rows, while the second index n increases with the number of the variables x_m. The example equation set can be solved for x_2:

$$3x_1 + x_2 = -2 \quad \leftrightarrow \quad x_2 = -2 - 3x_1$$
$$2x_1 + x_2 = 0 \quad \leftrightarrow \quad x_2 = -2x_1$$

which can be geometrically interpreted as shown in Figure 8. The solution is the point $x_1 = -2$ and $x_2 = -4$, where the vectors cross. You either get one solution, no solution, or an infinite number of solutions for any linear equation set. The general form, with m equations and n variables, is then:

$$
\begin{array}{ccccc}
a_{11}x_1 + a_{12}x_2 + & \cdots & +a_{1n}x_n & = & b_1 \\
a_{21}x_1 + a_{22}x_2 + & \cdots & +a_{2n}x_n & = & b_2 \\
& & \vdots & & \\
a_{m1}x_1 + a_{m2}x_2 + & \cdots & +a_{mn}x_n & = & b_m
\end{array} \tag{2.1}
$$

which can be written in a much denser form as:

$$Ax = b$$

with a matrix in a bold capital letter:

$$
A = \begin{pmatrix}
a_{11} & a_{12} & \cdots & a_{1n} \\
a_{21} & a_{22} & \cdots & a_{2n} \\
\vdots & \vdots & \ddots & \vdots \\
a_{m1} & a_{m2} & \cdots & a_{mn}
\end{pmatrix}
$$

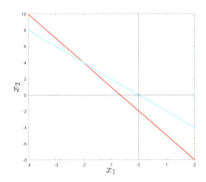

Figure 8. A linear equation system as vectors. The solution of the equation system is the cross-section of both vectors.

and vectors in bold lowercase letters:

$$
x = \begin{pmatrix} x_1 \\ x_2 \\ \vdots \\ x_n \end{pmatrix} \quad \text{and} \quad b = \begin{pmatrix} b_1 \\ b_2 \\ \vdots \\ b_m \end{pmatrix}.
$$

This is very convenient and compact. The solution is simply $x = A^{-1}b$. The exponent -1 indicates the matrix inverse and will be explained later. The matrix A basically describes n arrows with the arrow tail in the origin (zero-point), and the arrowhead on m coordinates in a space spanned by the coordinate system. The equation set 2.1 describes an **inhomogeneous system**. It becomes a **homogeneous system** if $b = 0$. If the system has a solution we have a **consistent system**, otherwise it is **inconsistent**. A linear equation system is also fully determined by the **augmented coefficient matrix**:

$$
(A \mid b) = \begin{pmatrix}
a_{11} & a_{12} & \cdots & a_{1n} & b_1 \\
a_{21} & a_{22} & \cdots & a_{2n} & b_2 \\
\vdots & \vdots & \ddots & \vdots & \vdots \\
a_{m1} & a_{m2} & \cdots & a_{mn} & b_m
\end{pmatrix}.
$$

YouTube: From an equation set to a matrix

▦ Simple matrix operations

Let's say we have a matrix B and multiply it by a matrix A from the left. What does it mean geometrically? It means that we transform the system B in a way that can be a rotation, a scaling, or any form of linear deformation of the space, which is spanned by the coordinate system of B. If the matrix B has p arrows pointing toward n coordinates within a n dimensional space, the multiplication of B by A from the left transforms the coordinate system from n dimensions to m dimensions. The final matrix C represents p arrows pointing to the new coordinates in a m dimensional space. Have in mind that the order of matrix multiplication matters in contrast to the multiplication with numbers. Because A transforms the coordinates of B from one space to another, the number of columns in A must equal the number of rows in B in order to have sufficient coupling of two spaces for the transformation or multiplication $AB = C$.

Division by a matrix does not exist, but division of a matrix by a scalar is possible.
YouTube: Multiplying matrices

Example 1: Multiply matrices

A more detailed scheme is:

where two c elements are calculated as follows:

$$c_{12} = a_{11}b_{12} + a_{12}b_{22}$$
$$c_{33} = a_{31}b_{13} + a_{32}b_{23}$$

What is the sum of two matrices geometrically? The matrix A has n arrows originating from the coordinate origin point to the m coordinates. If we sum up with matrix B, the n arrows of B start from the coordinates of A and land on the coordinates $A + B$. This is equivalent to having n arrows starting from the coordinate system origin and pointing to the coordinates described by $A + B$.
YouTube: Sum up matrices and scalar multiplication

Example 2: Sum and subtract matrices

The sum or subtraction of matrices with identical size is calculated entry-wise:

$$\begin{bmatrix} 1 & 3 & 1 \\ 1 & 0 & 0 \end{bmatrix} + \begin{bmatrix} 0 & 0 & 5 \\ 7 & 5 & 0 \end{bmatrix} = \begin{bmatrix} 1+0 & 3+0 & 1+5 \\ 1+7 & 0+5 & 0+0 \end{bmatrix}$$
$$= \begin{bmatrix} 1 & 3 & 6 \\ 8 & 5 & 0 \end{bmatrix}.$$

The scalar multiplication is also a transformation of the matrix. If you multiply a matrix by a number, you scale the matrix by this number without skewing or rotating it. The scalar multiplication by two doubles all coordinate values the arrows point to. Do not confuse this with scalar product, which is a form of inner product!
YouTube: Scalar multiplication

Example 3: Scalar multiplication

$$2 \cdot \begin{bmatrix} 1 & 8 & -3 \\ 4 & -2 & 5 \end{bmatrix} = \begin{bmatrix} 2 \cdot 1 & 2 \cdot 8 & 2 \cdot (-3) \\ 2 \cdot 4 & 2 \cdot (-2) & 2 \cdot 5 \end{bmatrix}$$
$$= \begin{bmatrix} 2 & 16 & -6 \\ 8 & -4 & 10 \end{bmatrix}$$

Another important operation is transposition (to interchange columns with rows).
YouTube: Transpose a matrix

Example 4: Transpose

$$M = \begin{pmatrix} 1 & 2 & 3 & 4 & 5 \\ 6 & 7 & 8 & 9 & 10 \\ 11 & 12 & 13 & 14 & 15 \\ 16 & 17 & 18 & 19 & 20 \end{pmatrix}$$

$$M^T = \begin{pmatrix} 1 & 6 & 11 & 16 \\ 2 & 7 & 12 & 17 \\ 3 & 8 & 13 & 18 \\ 4 & 9 & 14 & 19 \\ 5 & 10 & 15 & 20 \end{pmatrix}$$

■ **Square matrices**
Square matrices have as many rows as columns. Some square matrices are especially secure. One example is the matrix in diagonal Jordan form, which was the aim of the Gauss-Jordan method:

$$A = \begin{bmatrix} a_{11} & \cdots & 0 \\ \vdots & & \vdots \\ 0 & \cdots & a_{nn} \end{bmatrix}$$

Multiplying a diagonal matrix multiple times from the left is the same as using the number of multiplications as exponents of the diagonal elements:

$$A^3 B = \begin{bmatrix} a_{11} & \cdots & 0 \\ \vdots & & \vdots \\ 0 & \cdots & a_{nn} \end{bmatrix} \begin{bmatrix} a_{11} & \cdots & 0 \\ \vdots & & \vdots \\ 0 & \cdots & a_{nn} \end{bmatrix} \begin{bmatrix} a_{11} & \cdots & 0 \\ \vdots & & \vdots \\ 0 & \cdots & a_{nn} \end{bmatrix} B$$
$$= \begin{bmatrix} a_{11}^3 & \cdots & 0 \\ \vdots & & \vdots \\ 0 & \cdots & a_{nn}^3 \end{bmatrix} B$$

This is much better than multiplying non-diagonal matrices, which can be very frustrating after a while. A very important special form of the diagonal matrix is the identity matrix, with which has 1s as diagonal elements:

$$I =: \begin{bmatrix} 1 & \cdots & 0 \\ \vdots & & \vdots \\ 0 & \cdots & 1 \end{bmatrix}.$$

YouTube: Identity matrix

Also frequently mentioned is the zero matrix:

$$0 =: \begin{bmatrix} 0 & \cdots & 0 \\ \vdots & & \vdots \\ 0 & \cdots & 0 \end{bmatrix}.$$

One characteristic specifically of a square matrix is the trace. The trace is the sum of the diagonal elements:

$$\operatorname{tr}(A) = \sum_{i=1}^{n} a_{ii} = a_{11} + a_{22} + \cdots + a_{nn}$$

YouTube: See also the symmetric matrix

▮ Determinant

The determinant gives the area in a 2-by-2 matrix and the volume in a 3-by-3 matrix. What does it mean when the determinant is equal to zero for a 2-by-2 matrix? It means that the area of the matrix is equal to zero and therefore the vectors that compose the matrix are linear-dependent. In other words, it means that the vectors have parallel directions. From high school, we know that we can describe the position of any point y of a line in a function as an Origin O and a constant (c) that multiples a non-zero vector v $(y = c \cdot v + 0)$. For a plane and space, any point can be described as a linear combination of two independent vectors, respectively. Consequently, if the two vectors are linear-dependent, we are no longer able to describe any point in the plane, but only the points that are situated on a line that is parallel to the vectors. For 3-by-3, a determinant of zero indicates that at least 2 of the 3 vectors are linearly dependent and therefore only the point located on a plane can be described by this set of vectors. More generally, a matrix with a determinant of zero describes a transformation of the system that reduces its dimensions by 1. It is possible to collapse a system to lower the number of dimensions, but the opposite is not possible. Therefore, the inverse of matrix A with the determinant of A equals zero, which would geometrically result in an expansion of the system to a higher number of dimensions. This is not possible.

$$\det(A) = \begin{vmatrix} a & b \\ c & d \end{vmatrix} = ad - bc.$$

One possibility for calculating a determinant is via the initial reduction of the matrix. Larger matrices can be split into smaller matrices with the Laplace expansion:

$$\begin{vmatrix} a & b & c & d \\ e & f & g & h \\ i & j & k & l \\ m & n & o & p \end{vmatrix}$$

$$= +a \cdot \begin{vmatrix} f & g & h \\ j & k & l \\ n & o & p \end{vmatrix} - b \cdot \begin{vmatrix} e & g & h \\ i & k & l \\ m & o & p \end{vmatrix}$$

$$+ c \cdot \begin{vmatrix} e & f & h \\ i & j & l \\ m & n & p \end{vmatrix} - d \cdot \begin{vmatrix} e & f & g \\ i & j & k \\ m & n & o \end{vmatrix}$$

Please pay attention to the alternating signs $(+/-)$.

YouTube: Laplace expansion or cofactor expansion

Laplace expansion can be coupled with the Gauss method, as shown in Example 5.

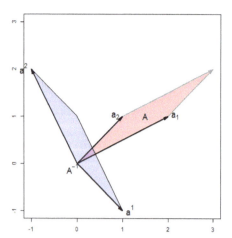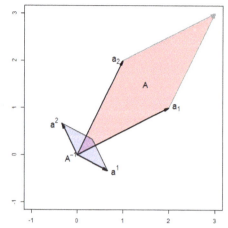

Figure 9. Geometrical interpretation of the matrix inverse and determinant impact. **Left:** $|A| = 1$. **Right:** $|A| = 3$ [4]. Copyright © 2021, Stack Exchange Inc, Licence: CC BY SA.

Example 5: Determinant after reduction with Gauss

$$
\det(A) = \begin{array}{c} \text{I)} \\ \text{II} - \text{I)} \\ \text{III} - 2\text{I)} \\ \text{IV} - 2\text{I)} \end{array} \begin{vmatrix} 1 & 3 & 2 & -6 \\ 1 & 2 & -2 & -5 \\ 2 & 4 & -2 & -9 \\ 2 & 4 & -6 & -9 \end{vmatrix} = \begin{vmatrix} 1 & 3 & 2 & -6 \\ 0 & -1 & -4 & 1 \\ 0 & -2 & -6 & 3 \\ 0 & -2 & -10 & 3 \end{vmatrix}
$$

The second row minus Row I, and Row III and IV minus 2 times the first row, gives the right matrix. Laplace expansion (see later) gives us the lower right matrix block. Further reduction and an additional isolation of the lower right block result in:

$$
\begin{array}{c} -\text{I)} \\ \text{II} - 2\text{I)} \\ \text{III} - 2\text{I)} \end{array} \begin{vmatrix} -1 & -4 & 1 \\ -2 & -6 & 3 \\ -2 & -10 & 3 \end{vmatrix} = \begin{vmatrix} 1 & 4 & -1 \\ 0 & 2 & 1 \\ 0 & -2 & 1 \end{vmatrix} = \begin{vmatrix} 2 & 1 \\ -2 & 1 \end{vmatrix}
$$
$$
= 2 \cdot 1 - 1 \cdot (-2) = \mathbf{4}
$$

A reduction is not always possible. The determinant of a 3-by-3 matrix can be obtained with the the rule of Sarrus,[1] where the first two columns can be written beside the determinant to facilitate the optical assessment of the diagonal product:

$$
\begin{vmatrix} a & b & c \\ d & e & f \\ g & h & i \end{vmatrix} \begin{matrix} a & b \\ d & e \\ g & h \end{matrix} \qquad \begin{vmatrix} a & b & c \\ d & e & f \\ g & h & i \end{vmatrix} \begin{matrix} a & b \\ d & e \\ g & h \end{matrix}
$$

$$
\begin{vmatrix} a & b & c \\ d & e & f \\ g & h & i \end{vmatrix} = \begin{array}{l} +(a \cdot e \cdot i + b \cdot f \cdot g + c \cdot d \cdot h) \\ -(c \cdot e \cdot g + a \cdot f \cdot h + b \cdot d \cdot i) \end{array}
$$

Determinants can also be solved efficiently with the **LU decomposition** explained in Example 10.
YouTube: The determinant

---

[1] French mathematician: Pierre Frédéric Sarrus (1798—1861).

Inversion

An invertible matrix A has linear-independent rows and columns. A matrix B is uniquely determined by A, if one has the symmetric and invertible matrix:

$$AB = BA = I$$

with identity matrix I.

Example 6: Inverse interpreted geometrically

Geometrically one can interpret the matrix [4]

$$A = \begin{bmatrix} 2 & 1 \\ 1 & 1 \end{bmatrix} \quad \text{with related vectors} \qquad \begin{aligned} a_1 &= \begin{bmatrix} 2 & 1 \end{bmatrix} \\ a_2 &= \begin{bmatrix} 1 & 1 \end{bmatrix} \end{aligned}$$

as an area in a 2-dimensional (2D) coordinate system shown in Figure 9 left in red. The shape of the inverse in blue is rotated by 90 degrees and is small in the directions where A is large. The vector a^2 is at right angles to a_1, and a^1 is at a right angle to a_2. The determinant $|A|$ describes the area in 2D and the volume in 3D. Here, the determinant $|A| = 1$ is one, and the area is preserved. If we change the matrix

$$A = \begin{bmatrix} 2 & 1 \\ 1 & 2 \end{bmatrix} \quad \text{with related vectors} \qquad \begin{aligned} a_1 &= \begin{bmatrix} 2 & 1 \end{bmatrix} \\ a_2 &= \begin{bmatrix} 1 & 2 \end{bmatrix} \end{aligned}$$

to get another determinant $|A| = 3$, we see that the area is changed.

We use **Cramer's rule**:

$$x_i = \frac{\det(A_i)}{\det(A)} = \frac{|A_i|}{|A|} \qquad i = 1, \ldots, n$$

with the site determinant $|A_i|$ to find the inverse analytically. The cofactors of the Laplace expansion can be saved in the so-called matrix of cofactors or comatrix C. Its transposed version C^T is the adjugate matrix:

$$A^{-1} = \frac{1}{|A|}C^T = \frac{1}{|A|}\operatorname{adj}(A) = \frac{1}{|A|}\begin{pmatrix} C_{11} & C_{21} & \cdots & C_{n1} \\ C_{12} & C_{22} & \cdots & C_{n2} \\ \vdots & \vdots & \ddots & \vdots \\ C_{1n} & C_{2n} & \cdots & C_{nn} \end{pmatrix}$$

Inversion of a 2 x 2 matrix

$$A^{-1} = \begin{bmatrix} a & b \\ c & d \end{bmatrix}^{-1} = \frac{1}{\det(A)}\begin{bmatrix} d & -b \\ -c & a \end{bmatrix} = \frac{1}{ad-bc}\begin{bmatrix} d & -b \\ -c & a \end{bmatrix}$$

Inversion of a 3 x 3 matrix

$$A = \begin{bmatrix} a & b & c \\ d & e & f \\ g & h & i \end{bmatrix} \qquad A^{-1} = \frac{1}{|A|}C^T$$

$$C = \begin{bmatrix} +\begin{vmatrix} e & f \\ h & i \end{vmatrix} & -\begin{vmatrix} d & f \\ g & i \end{vmatrix} & +\begin{vmatrix} d & e \\ g & h \end{vmatrix} \\[8pt] -\begin{vmatrix} b & c \\ h & i \end{vmatrix} & +\begin{vmatrix} a & c \\ g & i \end{vmatrix} & -\begin{vmatrix} a & b \\ g & h \end{vmatrix} \\[8pt] +\begin{vmatrix} b & c \\ e & f \end{vmatrix} & -\begin{vmatrix} a & c \\ d & f \end{vmatrix} & +\begin{vmatrix} a & b \\ d & e \end{vmatrix} \end{bmatrix}$$

$$C = \begin{bmatrix} +(ei-fh) & -(di-fg) & +(dh-eg) \\ -(bi-ch) & +(ai-cg) & -(ah-bg) \\ +(bf-ce) & -(af-cd) & +(ae-bd) \end{bmatrix}$$

$$C^T = \begin{bmatrix} +(ei-fh) & -(bi-ch) & +(bf-ce) \\ -(di-fg) & +(ai-cg) & -(af-cd) \\ +(dh-eg) & -(ah-bg) & +(ae-bd) \end{bmatrix}$$

YouTube: Inverse of a matrix
YouTube: Solve a linear equation set with the inverse of a Matrix

◼ The rank
Often, matrices derived from linear equation sets can be reduced, as not all equations are necessary to describe the system. We often search for the minimal matrix, which is described by the number of linear-independent rows or columns. Linear-independent rows and columns are sets of rows or columns where none of the rows or columns is a linear combination of the others (example of linear dependency: Row 1 equals to the sum of Rows 2, and Row 3 or Column 3 is three times Column 1. The number of linear-independent rows or columns is the rank of a matrix:

$$\operatorname{rank}(A) = \operatorname{rank}(A^T) = \operatorname{rk}(A)$$

This equation states that the rank computed on the rows is equal to the rank obtained on the columns and therefore, by definition, the rank of a matrix cannot be greater than the number of rows **and** columns in this matrix. In other words, if a matrix B has 6 rows and 2 columns, we can deduce that the rank is smaller than or equal to 2 as the rank cannot be greater than the number of columns. Consequently, a non-square matrix has by definition at least one linear-dependent column or row. For our matrix B, we know that we have at least 4 linear-dependent rows as we have 6 rows and the rank is smaller or equal to 2. The Gauss elimination is used determine the rank of a matrix by producing as many zeros as possible in the hope of removing as many **rows or columns** as possible, which results in the reduction of dimensions.
YouTube: Rank

◼ Solving a set of linear equation sets
Naive solving of a linear equation set can be very time-consuming.
YouTube: Naive solving of a linear equation set
We should start to check whether an equation set is solvable. The set of linear equations is solvable if the rank of the coefficient matrix A equals the rank of the augmented coefficient matrix $(A|b)$.

Example 7: Trivial solution only

$$2x_1 + x_2 = 0 \qquad \text{I}$$
$$x_1 - x_2 = 0 \qquad \text{II}$$

gives:

$$\text{II}: \qquad x_1 = x_2$$
$$\text{in I}: \qquad 2x_2 + x_2 = 0$$
$$x_1 = x_2 = 0$$

This is a trivial solution. Did we have a chance to find this out earlier?
We have a homogeneous system with as many equations m as variables n.

$$\operatorname{rk}\begin{bmatrix} 2 & 1 \\ 1 & -1 \end{bmatrix} = 2 = n$$

which indicates that we only have trivial solutions according to the Rouché-Capelli theorem. Because the determinant is also non-zero:

$$\det\begin{bmatrix} 2 & 1 \\ 1 & -1 \end{bmatrix} = -2 - 1 = -3 \neq 0$$

we would also not expect that the system is going to lose a dimension. Consequently, we expect only trivial solutions.

$$\text{rank}(A) = \text{rank}(A|b)$$

If the matrix is quadratic with $m = n$ and the determinant is not zero, then the set of linear equations is solvable. In a homogeneous system $Ax = 0$, it only leads to trivial solutions $a_{11} = a_{12} \ldots a_{mn} = 0$ or $x_1 = x_2 \ldots x_n = 0$. In an inhomogeneous system we get a unique solution. If $m < n$ the the rows or columns are linear-dependent. Thus, the related homogeneous system $Ax = 0$ has non-trivial solutions, which is good. If $m < n$, the inhomogeneous system has many non-unique solutions. The rules are summarized in the Rouché-Campelli theorem. We illustrate this with two examples, 7 & 8.

Example 8: Non-trivial solution

$$2x_1 + 2x_2 = 0 \qquad \text{I}$$
$$-x_1 - x_2 = 0 \qquad \text{II}$$

gives:

$$\text{II} \qquad\qquad x_1 = -x_2$$
$$\text{in I} \qquad -2x_2 + 2x_2 = 0$$

choose $x_2 = \lambda$ to get the solution:

$$\begin{pmatrix} x_1 \\ x_2 \end{pmatrix} = \begin{pmatrix} -\lambda \\ \lambda \end{pmatrix}$$

What would we expect? According to the Rouché-Capelli theorem we have many non-trivial solutions because we have a homogeneous system with fewer equations m than variables n:

$$\text{rk} \begin{bmatrix} 2 & 2 \\ -1 & -1 \end{bmatrix} = 1 < n$$

and the determinant also indicates a loss of one dimension:

$$\det \begin{bmatrix} 2 & 2 \\ -1 & -1 \end{bmatrix} = -2 + 2 = 0$$

A set of linear equations can be solved with the Rule of Cramer:

$$x_i = \frac{\det(A_i)}{\det(A)} = \frac{|A_i|}{|A|} \qquad i = 1, \ldots, n$$

using the site determinant $|A_i|$. The side determinant $|A_i|$ is the determinant $|A|$ where the i_{th} column is replaced by column vector b.

YouTube: Cramer's Rule — 2x2 and 3x3 matrices

Example 9: Cramer's Rule 2x2 matrix

Consider the linear system:

$$a_{11}x_1 + a_{12}x_2 = b_1$$
$$a_{21}x_1 + a_{22}x_2 = b_2$$

which in matrix format is:

$$\begin{bmatrix} a_{11} & a_{12} \\ a_{21} & a_{22} \end{bmatrix} \begin{bmatrix} x_1 \\ x_2 \end{bmatrix} = \begin{bmatrix} b_1 \\ b_2 \end{bmatrix}.$$

The **Cramer Rule** is then:

$$x_1 = \frac{\begin{vmatrix} b_1 & a_{12} \\ b_2 & a_{22} \end{vmatrix}}{\begin{vmatrix} a_{11} & a_{12} \\ a_{21} & a_{22} \end{vmatrix}} = \frac{b_1 a_{22} - a_{12} b_2}{a_{11} a_{22} - a_{12} a_{21}},$$

$$x_2 = \frac{\begin{vmatrix} a_{11} & b_1 \\ a_{21} & b_2 \end{vmatrix}}{\begin{vmatrix} a_{11} & a_{12} \\ a_{21} & a_{22} \end{vmatrix}} = \frac{a_{11} b_2 - b_1 a_{21}}{a_{11} a_{22} - a_{12} a_{21}}$$

The denominator (fraction below) is not allowed to become zero which explains why the determinant of A must be non-zero.

The Cramer Rule visualizes the resolvability but it takes too much effort to get the solution. The solving process is faster when one subtracts rows stepwise from each other to bring the augmented coefficient matrix into a certain form. Either in the **Gauss form** (**row-echelon form**):

$$(\,A \mid b\,) = \begin{pmatrix} \bullet & \bullet & \bullet & \bullet & \mid & \bullet \\ 0 & \bullet & \bullet & \bullet & \mid & \bullet \\ 0 & 0 & 0 & \bullet & \mid & \bullet \\ 0 & 0 & 0 & \bullet & \mid & \bullet \end{pmatrix}.$$

with as many zeros as possible in the lower left-corner, or in the **Gauss-Jordan form**:

$$(\,A \mid b\,) = \begin{pmatrix} \bullet & 0 & 0 & 0 & \mid & \bullet \\ 0 & \bullet & \bullet & 0 & \mid & \bullet \\ 0 & 0 & 0 & \bullet & \mid & \bullet \\ 0 & 0 & 0 & \bullet & \mid & \bullet \end{pmatrix}.$$

with as many zeros as possible in the lower-left and upper-right corner. If one eliminates the elements with the largest absolute value first, one reduces rounding mistakes. This is called pivoting. The row-echelon form becomes a **reduced row-echelon form** if every leading coefficient is 1 and is the only non-zero entry in its column:

$$(\,A \mid b\,) = \begin{pmatrix} 1 & \neq 1 & \neq 1 & \neq 1 & \mid & \bullet \\ 0 & 1 & \neq 1 & \neq 1 & \mid & \bullet \\ 0 & 0 & 0 & 1 & \mid & \bullet \\ 0 & 0 & 0 & 1 & \mid & \bullet \end{pmatrix}.$$

We have three types of elementary row (column) operations:

- Interchanging two rows (columns)

- Multiplying a row (column) by a real number

- Adding a multiple of one row (column) to another row (column).

If we bring the Gauss form back to the equation set form, we get:

$$
\begin{aligned}
a_{11}x_1 + a_{12}x_2 + \ \cdots \ + a_{1n}x_n &= b_1 \\
a_{22}x_2 + \ \cdots \ + a_{2n}x_n &= b_2 \\
\vdots \\
a_{mn}x_n &= b_m
\end{aligned}
$$

where we can get our x values by working from the bottom to the top. If it is in the Gauss-Jordan form, we can directly read the solution.

YouTube: Gauss and Gauss-Jordan elimination

Additional reading

In practice, software usually uses more efficient algorithms like the LU decomposition.[a] The LU decomposition dissects a matrix A into two factors: a lower triangular matrix L and an upper triangular matrix U:

$$A = LU$$

$$
\begin{bmatrix} a_{11} & a_{12} & a_{13} \\ a_{21} & a_{22} & a_{23} \\ a_{31} & a_{32} & a_{33} \end{bmatrix} = \begin{bmatrix} l_{11} & 0 & 0 \\ l_{21} & l_{22} & 0 \\ l_{31} & l_{32} & l_{33} \end{bmatrix} \begin{bmatrix} u_{11} & u_{12} & u_{13} \\ 0 & u_{22} & u_{23} \\ 0 & 0 & u_{33} \end{bmatrix}
$$

It resembles the Gauss elimination, whereby the upper triangular matrix can be compared with the Gauss form and the lower triangular matrix collects the steps necessary to get the Gauss form. The diagonal elements are usually set to one $l_{11} = l_{22} = l_{33} = 1$ for convenience and because we have more unknowns than equations. As you already saw in the Gauss elimination, it is practical to change row and column entries. Therefore, algorithms use a permutation matrix P to change the rows of matrix A and the permutation matrix Q to change the columns of matrix A. A is multiplied by P from the left and by Q from the right. A permutation matrix has n 1s and otherwise zero entries. Each row and each column has only a single 1. It could look like:

$$
P = \begin{pmatrix} 0 & 1 & 0 \\ 1 & 0 & 0 \\ 0 & 0 & 1 \end{pmatrix} \quad \text{or} \quad Q = \begin{pmatrix} 1 & 0 & 0 \\ 0 & 0 & 1 \\ 0 & 1 & 0 \end{pmatrix}
$$

The LU decomposition is then:

$$PAQ = LU$$

The solution of the system with linear equations can be obtained with setting the LU decomposition into the general equation set form:

$$Ax = b \ \Rightarrow LUx = b \ \Rightarrow Ly = b \text{ and } Ux = y$$

Solving first $Ly = b$ and then $Ux = y$ gives us x as shown in the following Example 10 and the video:

YouTube: LU decomposition

---

[a] Polish mathematician: Tadeusz Banachiewicz (1882—1954).

Example 10: LU decomposition of 2-by-2 matrix

We have the system:

$$\begin{bmatrix} 6 & 1 \\ 4 & 2 \end{bmatrix} \begin{bmatrix} x_1 \\ x_2 \end{bmatrix} = \begin{bmatrix} 1 \\ 2 \end{bmatrix}$$

We dissect the matrix A into the lower and upper triangular matrix:

$$\begin{bmatrix} 6 & 1 \\ 4 & 2 \end{bmatrix} = \begin{bmatrix} l_{11} & 0 \\ l_{21} & l_{22} \end{bmatrix} \begin{bmatrix} u_{11} & u_{12} \\ 0 & u_{22} \end{bmatrix}.$$

We obtain the original matrix if we multiply the lower with the upper matrix. We obtain 4 equations with 6 unknowns:

$$l_{11} \cdot u_{11} + 0 \cdot 0 = 6$$
$$l_{11} \cdot u_{12} + 0 \cdot u_{22} = 1$$
$$l_{21} \cdot u_{11} + l_{22} \cdot 0 = 4$$
$$l_{21} \cdot u_{12} + l_{22} \cdot u_{22} = 2$$

This underdetermined system allows us to set 2 variables to arbitrary non-zero values such as 1 for the diagonal elements of the lower matrix: $l_{11} = l_{22} = 1$. This is up to you to decide. We would get infinitely many ways as there are also many ways to perform a Gauss elimination. We need only 1 convenient LU decomposition. The following matrix entries are then:

$$u_{11} = 6, \quad u_{12} = 1, \quad l_{21} = \frac{2}{3}, \quad u_{22} = \frac{4}{3}.$$

Substituting these values into the LU decomposition above yields:

$$\begin{bmatrix} 6 & 1 \\ 4 & 2 \end{bmatrix} = \begin{bmatrix} 1 & 0 \\ \frac{2}{3} & 1 \end{bmatrix} \begin{bmatrix} 6 & 1 \\ 0 & \frac{4}{3} \end{bmatrix} = LU.$$

Solving $Ly = b$ gives:

$$\begin{bmatrix} 1 & 0 \\ \frac{2}{3} & 1 \end{bmatrix} \begin{bmatrix} y_1 \\ y_2 \end{bmatrix} = \begin{bmatrix} 1 \\ 2 \end{bmatrix}$$

and thus $y_1 = 1$ and $y_2 = \frac{4}{3}$. Solving $Ux = y$

$$\begin{bmatrix} 6 & 1 \\ 0 & \frac{4}{3} \end{bmatrix} \begin{bmatrix} x_1 \\ x_2 \end{bmatrix} = \begin{bmatrix} 1 \\ \frac{4}{3} \end{bmatrix}$$

gives us the solution $x_2 = 1$ and $x_1 = 0$. Let's check whether the solution is correct:

$$\begin{bmatrix} 6 & 1 \\ 4 & 2 \end{bmatrix} \begin{bmatrix} 0 \\ 1 \end{bmatrix} = \begin{bmatrix} 1 \\ 2 \end{bmatrix}$$

Yes, it works.

Extra:

In the Gauss elimination, we would multiply the first row by $\frac{2}{3}$ (from L) and subtract this from the second row:

$$\begin{bmatrix} 6 & 1 \\ 4 - \frac{2}{3} \cdot 6 & 2 - \frac{2}{3} \cdot 1 \end{bmatrix} = U = \begin{bmatrix} 6 & 1 \\ 0 & \frac{4}{3} \end{bmatrix}$$

Determinant:

The determinant is easier to solve with the LU decomposition because it depends only on the diagonal elements. The determinant is $\det(A) = \det(L) \det(U)$. The determinant of the matrix $\det(A) = (1 \cdot 1) \cdot (l_{11} l_{22}) \cdot (u_{11} u_{22}) = 1 \cdot 1 \cdot (\frac{4}{3} \cdot 6) = 8$ is only dependent on the diagonal entries.

The storage demand for the lower and upper coefficient matrix can be huge. Thus one can make use of many zero values and use iterative procedures with estimators. An example is the Gauss–Seidel method,[2] which needs much less storage but cannot be used for parallel computing. For matrices larger than $n = 1000$, the Strassen algorithm[3] might be faster with lower numeric stability. Only the Coppersmith–Winograd algorithm[4] is even faster but only for matrices too big for modern computers.

■ Eigenvectors and Eigenvalues

Modeling complicated systems can result in very ugly mathematical problems. Instead of solving these kinds of problems, mathematicians transform the space until the problem becomes easier to solve. The solution of the nice, beautiful system can then be transformed back to the original space. Imagine, you are sitting in your room, and you damn the authors of this script, the expectations, and math in general. Now, you squeeze and stretch the dimensions of your room until this sheet of paper becomes infinitely small and your head so big that you master its challenges without any problem. After you are done, you multiply your entire existence with the inverse of the transformation matrix and find a fully understood and solved script in front of you. How complicated a problem is also depends on psychology. Forget everything around you; allow yourself to make mistakes and to be a child on discovery. The exam is not important, but the excitement to learn and the willingness to improve is. This is my favorite psychologic transformation matrix. Give it a try. So why is this imagination important for the Eigenvectors? **Eigenvectors are the directions in space, which are not deformed during transformation**. Imagine, everything gets squeezed and

---

[2] German mathematician: Philipp Ludwig von Seidel (1821—1896).
[3] German mathematician Volker Strassen (1936—today).
[4] Israeli American computer scientist Shmuel Winograd (1936—today) and American mathematician Don Coppersmith (1950—today).

changes—just not in the directions of these Eigenvectors. The directions along the Eigenvectors become merely scaled by the Eigenvalue. Would it not be wonderful to see a problem related to these vectors, compared to an original unfavorable coordinate system? But how can we find out which directions are not influenced by a transformation? Let's say we have a transformation matrix A and look for a vector v that can only be scaled by a factor but not deformed by a matrix. Then we could just say that the matrix is only like a factor. Let us call it λ and imagine the **Eigenspace**:

$$Av = \lambda v$$

The idea is not bad, but the right side does not have the same structure as the left one. The identity matrix helps:

$$(A - \lambda I)v = 0$$

The matrix $(A - \lambda I)$ should now be singular, to get non-zero solutions for the Eigenvector v. The Eigenvector $v = 0$ would be trivial and useless. A matrix is singular if its determinant is zero:

$$\det(A - \lambda I) = 0,$$

which also means that the matrix is not invertible. We are interested in vectors, which are scaled but not deformed by a transformation matrix A multiplied by the left. We calculate first the Eigenvalues and than the Eigenvectors.

Example 11: First Eigenvalues than Eigenvectors

We take the matrix:

$$\begin{pmatrix} 3 & 6 \\ 1 & 4 \end{pmatrix}$$

and get the Eigenvalues by:

$$A - \lambda I = \begin{bmatrix} 3 - \lambda & 6 \\ 1 & 4 - \lambda \end{bmatrix}$$

$$\det(A - \lambda I) = 0 = (3 - \lambda)(4 - \lambda) - 6$$

$$0 = \lambda^2 - 7\lambda + 6$$

$$0 = \lambda^2 + p\lambda + q \quad \text{(p-q-equation)}$$

$$\lambda_{1,2} = -\frac{p}{2} \pm \sqrt{\left(\frac{p}{2}\right)^2 - q}$$

$$\lambda_{1,2} = \frac{7}{2} \pm \sqrt{\left(\frac{-7}{2}\right)^2 - 6}$$

$$\lambda_{1,2} = \frac{7}{2} \pm \frac{5}{2}$$

$$x_1 = 1$$

$$x_2 = 6$$

$$0 = (\lambda - 6)(\lambda - 1).$$

The term $\det(A - \lambda I)$ creates a polynomial which we call a **characteristic polynomial** with maximal n solutions for a $n \times n$ matrix. Setting it to zero causes it be renamed a **characteristic equation**. Now we have the Eigenvalues $\lambda_1 = 6$ and $\lambda_2 = 1$ and search the related Eigenvectors starting with $\lambda_1 = 6$:

$$(A - \lambda I)v = 0$$

$$\left(\begin{bmatrix} 3 & 6 \\ 1 & 4 \end{bmatrix} - \begin{bmatrix} 6 & 0 \\ 0 & 6 \end{bmatrix} \right) v = 0$$

$$\begin{bmatrix} -3 & 6 \\ 1 & -2 \end{bmatrix} v = 0$$

to obtain the Eigenvector $v_1 = \begin{bmatrix} 2 \\ 1 \end{bmatrix}$ directly. We can also write it in more detail:

$$-3v_a + 6v_b = 0$$

$$v_a - 2v_b = 0$$

resulting in $v_a = 2v_b$ for the first equation. The second equation gives $2v_b - 2v_b = 0$ which means that v_b can be everything but the equation set is still solved. We have infinitely many solutions. So we can set $v_a = 2$ with which we know that v_b must be one $v_b = 1$. We obtain the non-unique solution $v_1 = \begin{bmatrix} 2 \\ 1 \end{bmatrix}$. The reason is that we can only stretch the Eigenvectors.

We proceed with the second Eigenvalue $\lambda_2 = 1$:

$$(A - \lambda I)v = 0$$

$$\left(\begin{bmatrix} 3 & 6 \\ 1 & 4 \end{bmatrix} - \begin{bmatrix} 1 & 0 \\ 0 & 1 \end{bmatrix} \right) v = 0$$

$$\begin{bmatrix} 2 & 6 \\ 1 & 3 \end{bmatrix} v = 0$$

to obtain Eigenvector $v_2 = \begin{bmatrix} -3 \\ 1 \end{bmatrix}$.

Summary:

1. Use the determinant of $(A - \lambda I)$ to get a polynomial of degree n.

2. Find the Eigenvalues by identifying the roots of the characteristic equation with $\det(A - \lambda I) = 0$.

3. Find to each Eigenvalue the associated Eigenvector via $(A - \lambda I)v = 0$.

If all Eigenvalues have another value, we call them **simple** and the associated Eigenvectors are independent. If an Eigenvalue appears several times, we say that the Eigenvalue has **multiplicity** k. Software programs like MATLAB return normalized Eigenvectors with unit length.

YouTube: Eigenvalues and Eigenvectors

Fast equation for 2-by-2 matrix
For the dimension 2 to 4, we have fast equations to solve the Eigenvalues. From the matrix:

$$\begin{bmatrix} a & b \\ c & d \end{bmatrix}$$

we obtain the characteristic polynomial:

$$\det \begin{bmatrix} \lambda - a & -b \\ -c & \lambda - d \end{bmatrix} = \lambda^2 - (a+d)\lambda + (ad - bc)$$

$$= \lambda^2 - \lambda \operatorname{tr}(A) + \det(A)$$

from which we get the Eigenvalues with the equation:

$$\lambda = \frac{\operatorname{tr}(A) \pm \sqrt{\operatorname{tr}^2(A) - 4\det(A)}}{2}.$$

with the distance between Eigenvalues:

$$\Delta =: \sqrt{\operatorname{tr}^2(A) - 4\det(A)}$$

which will play a role in the stability theory in Block 3. Additional information: for a matrix of higher dimension, it already seems to be more complicated.

$$\det(\alpha I - A) = 0$$

$$= \alpha^3 - \alpha^2 \operatorname{tr}(A) - \alpha \frac{1}{2}\left(\operatorname{tr}(A^2) - \operatorname{tr}^2(A)\right) - \det(A)$$

and for Dimension 4 it is not considered overly useful.

Example 12: Get Eigenvalue via fast equations

$$A = \begin{bmatrix} 4 & 3 \\ -2 & -3 \end{bmatrix}$$

gives $\operatorname{tr}(A) = 4 - 3 = 1$, $\det(A) = 4(-3) - 3(-2) = -6$ and the characteristic equation:

$$0 = \lambda^2 - \lambda - 6 = (\lambda - 3)(\lambda + 2)$$

with Eigenvalues 3 and -2. The Eigenvectors are worked out using:

$$A - 3I = \begin{bmatrix} 1 & 3 \\ -2 & -6 \end{bmatrix}, \qquad A + 2I = \begin{bmatrix} 6 & 3 \\ -2 & -1 \end{bmatrix}$$

$(3, -1)$ for Eigenvalue 3 and $(1, -2)$ for Eigenvalue -2.

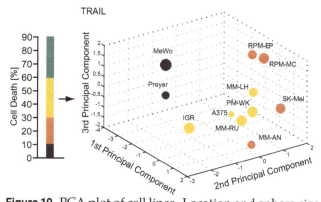

Figure 10. PCA plot of cell lines. Location and sphere size indicate which cell lines behave more similar than others. Cell lines which are close behave similarly. Source: [5]. Copyright © 2013, Macmillan Publishers Limited.

■ **More on biological data as a matrix**
Often, the network topology is unknown (but of course existent). Detected molecules change with time, and their behavior depends on their connection to other elements. Among them, interlinked elements correlate or anti-correlate with each other. The more different perturbations are applied, the more likely a valid connection can be detected. This is a property which can be used for, *e.g.*, reconstruction of networks and for data-driven modeling. The simultaneous observation of more than one outcome variable is the definition of **multivariate statistics**. Related data-tables with multiple variables are actually nothing other than matrices with molecular features as rows and experimental datasets, conditions, or organisms as columns:

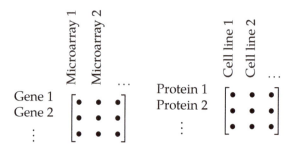

The columns represent the dimensions or coordinate axes of the data and the rows contain coordinate values on the given coordinate system. Consequently, each gene is a vector in a coordinate system spanned by the matrix columns.

To illustrate the application of matrix calculation, we will briefly discuss here the statistical method of *Principal Component Analysis* (PCA): sometimes, we want to know whether the columns, with all the row entries projecting to them, have subsets that are more similar to one another than to other data points. Now, imagine a dataset with 40 coordinate axes representing 40 microarrays, and then try to place tens of thousands of points representing genes into this coordinate system. Horrible! But maybe we can find another coordinate system which is visually more inviting and more informative. Let's say 40 data points representing microarrays in a 3-dimensional coordinate system, where the biggest differences between the microarrays are selected and are thus pronounced. The variability in each row causes the differences between the microarrays. Data analysts use **Principal Component Analysis** (**PCA**) to do exactly this. One starts with 1 additional coordinate axis which explains the most variability in the old dataset. This is the first **principal component**. The algorithm then searches for another coordinate axis which is orthogonal (90°) to the first principal component and again describes as much variability as possible. This procedure goes on until all genes project to the new coordinate system. After one has all these principle components, one neglects stepwise the components representing the lowest variability until 2 to 4 remain. Ideally, the remaining principal components explain more than 80% of the variability. Principal Component Analysis is used, *e.g.*, as standard quality control for microarrays. Are the treated and the untreated samples in two separate groups? Is one replicate completely different from the others and might it represent an outlier? As an example, PCA also helps us to understand the microbiome better. Are different bacteria types more closely related to others and how does it change if they are exposed to drugs or another diet? A collection of micro-organisms might build one cluster. After the diet is changed, the micro-organisms might be found in another cluster. Micro-organisms which have not changed might not be affected by the diet change. In Figure 10, you see an example with different cell lines and their response to TNF-related apoptosis-inducing ligand (TRAIL) [5]. The dataset contains the base level of 17 core apoptosis proteins in 11 melanoma cell lines under different conditions with 612 measurements in total. The authors used network information to group proteins to functional network motifs, which allowed a higher accuracy in estimating the apoptosis-inducing impact of drugs.

YouTube: PCA step for step

Additional reading

The PCA is related to the **partial least square regression** (**PLSR**). Here, we separate our data matrix in two parts, whereby the upper part represents phenomenological readouts such as viability:

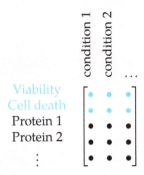

Both blocks are subject to a dimension reduction approach but now with the aim to maximize the co-variance between both blocks. PCA uses the variance and PLSR uses co-variance. Such an approach is, *e.g.*, used for systematic drug testing and for initial hints for network modeling. An example is shown in Figure 11. The authors created an high-dimensional data block with 5 different RAF/MEK inhibitors, 7 doses, 5 time points, 21 protein levels, and 10 other cell lines [6]. Here, drug dose is the first and drug type is the second principal component. The adjusted variable importance in the projection (VIP) explains which protein had the most prominent negative or positive impact on the cell viability at which time. This led to the identification of a consistent down-regulation of the JNK/c-Jun pathway upon RAF/MEK inhibitor treatment at early time points, but an up-regulation of 6 cell lines at later time points. In 4 out of 10 cell lines, JNK/c-Jun up-regulation caused a subset of cells to become quiescent and apoptosis-resistant [6].

We have learned that matrices are used in several applications in biology. Matrices can represent biochemical networks and data caused by them. In the following section, we will learn how to work with matrices. These basics are fundamental for the remainder of the book and very useful for your career, independent of whether you want to mainly work with the keyboard or the pipette in the future. It will help us to analyze and understand what we do.

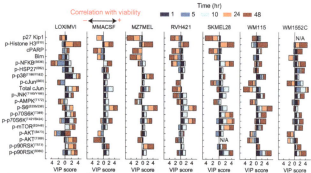

Figure 11. After the perturbation of the system with different conditions, we see which protein on the left correlates positively or negatively with the cell variability in different cell lines such as LOXIMVI or SKMEL28. The cell lines behave quite differently. The importance of each player depends on the time, so phosphorylated Histone H3 is first correlated with the cell viability after 24h or 48h. Figure source (cropped): [6], Licence: CC BY 4.0.

References

[1] Robert A Weinberg. Coming full circle—From endless complexity to simplicity and back again. *Cell*, 157(1):267–271, 2014. https://doi.org/10.1016/j.cell.2014.03.004.

[2] Olaf Wolkenhauer. Why model? *Frontiers in Physiology*, 5:21, 2014. https://doi.org/10.3389/fphys.2014.00021.

[3] Damian Szklarczyk, Annika L Gable, Katerina C Nastou, David Lyon, Rebecca Kirsch, Sampo Pyysalo, Nadezhda T Doncheva, Marc Legeay, Tao Fang, Peer Bork, Lars J Jensen, and Christian von Mering. The STRING database in 2021: customizable protein-protein networks, and functional characterization of user-uploaded gene/measurement sets. *Nucleic Acids Res.*, 49(D1), 2021. https://doi.org/10.1093/nar/gkaa1074.

[4] math.stackexchange.com/questions/295250/geometric-interpretations-of-matrix-inverses. December 2022.

[5] Egle Passante, Maximilian L Würstle, Christian T Hellwig, Martin Leverkus, and Markus Rehm. Systems analysis of apoptosis protein expression allows the case-specific prediction of cell death responsiveness of melanoma cells. *Cell Death and Differentiation*, 20(11):1521, 2013. https://doi.org/10.1038/cdd.2013.106.

[6] Mohammad Fallahi-Sichani, Nathan J Moerke, Mario Niepel, Tinghu Zhang, Nathanael S Gray, and Peter K Sorger. Systematic analysis of brafv600e melanomas reveals a role for jnk/c-jun pathway in adaptive resistance to drug-induced apoptosis. *Molecular Systems Biology*, 11(3):797, 2015. https://doi.org/10.15252/msb.20145877.

3. Exercises

■ Representation

Write the following systems of equation in matrix form:

$$y_1 = 2x_1 + 3x_2$$
$$y_2 = x_2 - 5x_1 \qquad (3.1)$$

$$\dot{x} = 1 - x - y$$
$$\dot{y} = 1 + x - y \qquad (3.2)$$

$$y_1 = \delta_{11}x_1 + \delta_{12}x_2 + \delta_{13}x_3$$
$$y_2 = \delta_{22}x_2 + \delta_{21}x_1 + \delta_{23}x_3 \qquad (3.3)$$
$$y_3 = \delta_{33}x_3 + \delta_{32}x_2 + \delta_{31}x_1$$

Write the following matrix equation as a system of differential equations:

$$\frac{d}{dt}\begin{pmatrix} x_1 \\ x_2 \end{pmatrix} = \begin{pmatrix} 3 & -2 \\ 1 & 5 \end{pmatrix}\begin{pmatrix} x_1 \\ x_2 \end{pmatrix} + \begin{pmatrix} 1 \\ 0 \end{pmatrix}u \qquad (3.4)$$

■ Basic operations on matrices

Perform the following calculations:

$$\begin{pmatrix} 3 & -2 \\ 1 & 5 \end{pmatrix} + \begin{pmatrix} 4 & -4 \\ 3 & 0 \end{pmatrix} = \qquad (3.5)$$

$$\begin{pmatrix} 3 & -2 \\ 1 & 5 \end{pmatrix} - \begin{pmatrix} 4 & -4 \\ 3 & 0 \end{pmatrix} = \qquad (3.6)$$

$$5\begin{pmatrix} 3 & -2 \\ 1 & 5 \end{pmatrix} = \qquad (3.7)$$

$$\begin{pmatrix} 3 & -2 \\ 1 & 5 \end{pmatrix}\begin{pmatrix} 4 & -4 \\ 3 & 0 \end{pmatrix} = \qquad (3.8)$$

$$\begin{pmatrix} 4 & -4 \\ 3 & 0 \end{pmatrix}\begin{pmatrix} 3 & -2 \\ 1 & 5 \end{pmatrix} = \qquad (3.9)$$

Transpose the following matrices:

$$\begin{pmatrix} 3 & -2 \\ 1 & 5 \end{pmatrix}^T = \qquad (3.10)$$

$$\begin{pmatrix} 3 & -2 & 1 \\ 1 & 5 & 0 \\ 2 & -1 & 7 \end{pmatrix}^T = \qquad (3.11)$$

■ Determinant

Find the determinants of the following matrices:

$$\begin{pmatrix} 3 & -2 \\ 1 & 5 \end{pmatrix} \qquad (3.12)$$

$$\begin{pmatrix} 3 & -2 & 1 \\ 1 & 5 & 0 \\ 2 & -1 & 7 \end{pmatrix} \qquad (3.13)$$

$$\begin{pmatrix} 3 & -2 & 2 \\ 1 & 5 & 0 \\ 2 & -1 & 0 \end{pmatrix} \qquad (3.14)$$

$$\begin{pmatrix} 3 & -2 & 2 & 2 \\ 1 & 5 & 1 & 2 \\ 2 & -1 & -1 & -2 \\ 1 & 2 & 3 & 1 \end{pmatrix} \qquad (3.15)$$

■ Rank and inversion

Determine the rank of:

$$\begin{pmatrix} 3 & -2 \\ -6 & 4 \end{pmatrix} \qquad (3.16)$$

$$\begin{pmatrix} 3 & -2 \\ 6 & -3 \end{pmatrix} \qquad (3.17)$$

$$\begin{pmatrix} 3 & -2 & 1 \\ 1 & 5 & 0 \\ 1 & 2 & 3 \end{pmatrix} \qquad (3.18)$$

$$\begin{pmatrix} 1 & 1 & -3 \\ 1 & -1 & 2 \\ 2 & 0 & -1 \end{pmatrix} \qquad (3.19)$$

Invert the following matrix:

$$\begin{pmatrix} 3 & -2 \\ 6 & 3 \end{pmatrix} \qquad (3.20)$$

$$\begin{pmatrix} 2 & 1 & -1 \\ 0 & 2 & 1 \\ 5 & 2 & -3 \end{pmatrix} \qquad (3.21)$$

Linear systems of equations

Solve the following systems of equations ($Ax = b$). Consider thereby the rank of the coefficient matrix A and of the augmented coefficient matrix $(A|b)$.

$$
\begin{aligned}
2x_1 + x_2 - x_3 &= 0 \\
x_2 + x_3 &= 0
\end{aligned}
\tag{3.22}
$$

$$
\begin{aligned}
2x_1 + x_2 &= 0 \\
x_1 - x_2 &= 0
\end{aligned}
\tag{3.23}
$$

$$
\begin{aligned}
x_1 + 2x_2 &= 1 \\
x_1 + 2x_2 &= 2
\end{aligned}
\tag{3.24}
$$

$$
\begin{aligned}
5x_1 + x_2 &= 2 \\
x_1 - 2x_2 &= 7
\end{aligned}
\tag{3.25}
$$

$$
\begin{aligned}
2x_1 + x_2 - x_3 &= -5 \\
x_2 + x_3 &= 1
\end{aligned}
\tag{3.26}
$$

Notes

4. Solutions

Do not betray yourself!

Exercises

▇ Representation

Task 3.1

$$\begin{aligned} y_1 &= 2x_1 + 3x_2 \\ y_2 &= x_2 - 5x_1 \end{aligned} \quad \leftrightarrow \quad \begin{aligned} 2x_1 + 3x_2 &= y_1 \\ -5x_1 + x_2 &= y_2 \end{aligned}$$

$$\begin{pmatrix} 2 & 3 \\ -5 & 1 \end{pmatrix} \begin{pmatrix} x_1 \\ x_2 \end{pmatrix} = \begin{pmatrix} y_1 \\ y_2 \end{pmatrix}$$

Task 3.2

$$\begin{aligned} \dot{x} &= 1 - x - y \\ \dot{y} &= 1 + x - y \end{aligned} \quad \leftrightarrow \quad \begin{pmatrix} -1 & -1 \\ 1 & -1 \end{pmatrix} \begin{pmatrix} x \\ y \end{pmatrix} + \begin{pmatrix} 1 \\ 1 \end{pmatrix} = \begin{pmatrix} \dot{x} \\ \dot{y} \end{pmatrix}$$

Task 3.3

$$\begin{aligned} \dot{y}_1 &= \delta_{11}x_1 + \delta_{12}x_2 + \delta_{13}x_3 \\ \dot{y}_2 &= \delta_{22}x_2 + \delta_{21}x_1 + \delta_{23}x_3 \\ \dot{y}_3 &= \delta_{33}x_3 + \delta_{32}x_2 + \delta_{31}x_1 \end{aligned}$$

$$\begin{pmatrix} \dot{y}_1 \\ \dot{y}_2 \\ \dot{y}_3 \end{pmatrix} = \begin{pmatrix} \delta_{11} & \delta_{12} & \delta_{13} \\ \delta_{21} & \delta_{22} & \delta_{23} \\ \delta_{31} & \delta_{32} & \delta_{33} \end{pmatrix} \begin{pmatrix} x_1 \\ x_2 \\ x_3 \end{pmatrix}$$

Task 3.4

$$\frac{d}{dt} \begin{pmatrix} x_1 \\ x_2 \end{pmatrix} = \begin{pmatrix} 3 & -2 \\ 1 & 5 \end{pmatrix} \begin{pmatrix} x_1 \\ x_2 \end{pmatrix} + \begin{pmatrix} 1 \\ 0 \end{pmatrix} u$$

$$\frac{dx_1}{dt} = 3x_1 - 2x_2 + u$$

$$\frac{dx_2}{dt} = x_1 + 5x_2$$

▇ Basic operations on matrices

Task 3.5

$$\begin{pmatrix} 3 & -2 \\ 1 & 5 \end{pmatrix} + \begin{pmatrix} 4 & -4 \\ 3 & 0 \end{pmatrix} = \begin{pmatrix} 7 & -6 \\ 4 & 5 \end{pmatrix}$$

Task 3.6

$$\begin{pmatrix} 3 & -2 \\ 1 & 5 \end{pmatrix} - \begin{pmatrix} 4 & -4 \\ 3 & 0 \end{pmatrix} = \begin{pmatrix} -1 & 2 \\ -2 & 5 \end{pmatrix}$$

Task 3.7

$$5 \begin{pmatrix} 3 & -2 \\ 1 & 5 \end{pmatrix} = \begin{pmatrix} 15 & -10 \\ 5 & 25 \end{pmatrix}$$

Task 3.8

$$\begin{pmatrix} 3 & -2 \\ 1 & 5 \end{pmatrix} \begin{pmatrix} 4 & -4 \\ 3 & 0 \end{pmatrix} =$$

$$\begin{pmatrix} (3 \cdot 4) + (-2 \cdot 3) & (3 \cdot (-4)) + (-2 \cdot 0) \\ (1 \cdot 4) + (5 \cdot 3) & (1 \cdot (-4)) + (5 \cdot 0) \end{pmatrix} = \begin{pmatrix} 6 & -12 \\ 19 & -4 \end{pmatrix}$$

Task 3.9

$$\begin{pmatrix} 4 & -4 \\ 3 & 0 \end{pmatrix} \begin{pmatrix} 3 & -2 \\ 1 & 5 \end{pmatrix} =$$

$$\begin{pmatrix} 12 - 4 & -8 - 20 \\ 9 + 0 & -6 + 0 \end{pmatrix} = \begin{pmatrix} 8 & -28 \\ 9 & -6 \end{pmatrix}$$

Task 3.10

$$\begin{pmatrix} 3 & -2 \\ 1 & 5 \end{pmatrix}^T = \begin{pmatrix} 3 & 1 \\ -2 & 5 \end{pmatrix}$$

Task 3.11

$$\begin{pmatrix} 3 & -2 & 1 \\ 1 & 5 & 0 \\ 2 & -1 & 7 \end{pmatrix}^T = \begin{pmatrix} 3 & 1 & 2 \\ -2 & 5 & -1 \\ 1 & 0 & 7 \end{pmatrix}$$

▇ Determinant

Task 3.12

$$\det \begin{pmatrix} 3 & -2 \\ 1 & 5 \end{pmatrix} = \begin{vmatrix} 3 & -2 \\ 1 & 5 \end{vmatrix} = (3 \cdot 5) - (1 \cdot (-2)) = 17$$

Task 3.13

$$\det \begin{pmatrix} 3 & -2 & 1 \\ 1 & 5 & 0 \\ 2 & -1 & 7 \end{pmatrix} = (3 \cdot 5 \cdot 7) + (-2 \cdot 0 \cdot 2) + (1 \cdot 1 \cdot (-1))$$

$$- (2 \cdot 5 \cdot 1) - (-1 \cdot 0 \cdot 3) - (7 \cdot 1 \cdot (-2))$$

$$= 108$$

Task 3.14

$$\det \begin{pmatrix} 3 & -2 & 2 \\ 1 & 5 & 0 \\ 2 & -1 & 0 \end{pmatrix} = 0 + 0 - 2 - 20 - 0 - 0 = -22 \quad (4.1)$$

Task 3.15

$$\begin{vmatrix} 3 & -2 & 2 & 2 \\ 1 & 5 & 1 & 2 \\ 2 & -1 & -1 & -2 \\ 1 & 2 & 3 & 1 \end{vmatrix}$$

$$= 3 \begin{vmatrix} 5 & 1 & 2 \\ -1 & -1 & -2 \\ 2 & 3 & 1 \end{vmatrix} - (-2) \begin{vmatrix} 1 & 1 & 2 \\ 2 & -1 & -2 \\ 1 & 3 & 1 \end{vmatrix}$$

$$+ 2 \begin{vmatrix} 1 & 5 & 2 \\ 2 & -1 & -2 \\ 1 & 2 & 1 \end{vmatrix} - 2 \begin{vmatrix} 1 & 5 & 1 \\ 2 & -1 & -1 \\ 1 & 2 & 3 \end{vmatrix}$$

$$= 3(-5 - 4 - 6 + 4 + 30 + 1)$$

$$+ 2(-1 - 2 + 12 + 2 + 6 - 2)$$

$$+ 2(-1 - 10 + 8 + 2 + 4 - 10)$$

$$- 2(-3 - 5 + 4 + 1 + 2 - 30)$$

$$= 138$$

■ Rank and inversion

Task 3.16

$$\begin{pmatrix} 3 & -2 \\ -6 & 4 \end{pmatrix} \begin{array}{c} \text{II} + 2 \cdot \text{I} \\ = \end{array} \begin{pmatrix} 3 & -2 \\ 0 & 0 \end{pmatrix} \rightarrow \text{Rank is 1}$$

Task 3.17

$$\begin{pmatrix} 3 & -2 \\ 6 & -3 \end{pmatrix} \begin{array}{c} \text{II} - 2 \cdot \text{I} \\ = \end{array} \begin{pmatrix} 3 & -2 \\ 0 & 1 \end{pmatrix} \rightarrow \text{Rank is 2}$$

Task 3.18

$$\begin{pmatrix} 3 & -2 & 1 \\ 1 & 5 & 0 \\ 1 & 2 & 3 \end{pmatrix} \begin{array}{c} \text{I} - 3 \cdot \text{II} \\ = \\ \text{III} - \text{II} \end{array} \begin{pmatrix} 0 & -17 & 1 \\ 1 & 5 & 0 \\ 0 & -3 & 3 \end{pmatrix} \rightarrow \text{Rank is 3}$$

Task 3.19

$$\begin{pmatrix} 1 & 1 & -3 \\ 1 & -1 & 2 \\ 2 & 0 & -1 \end{pmatrix} \begin{array}{c} \text{II} - \text{I} \\ = \\ \text{III} - 2 \cdot \text{I} \end{array} \begin{pmatrix} 1 & 1 & -3 \\ 0 & -2 & 5 \\ 0 & -2 & 5 \end{pmatrix}$$

$$\begin{array}{c} \text{III} - \text{II} \\ = \end{array} \begin{pmatrix} 1 & 1 & -3 \\ 0 & -2 & 5 \\ 0 & 0 & 0 \end{pmatrix} \rightarrow \text{Rank is 2}$$

Task 3.20

$$\begin{vmatrix} 3 & -2 \\ 6 & 3 \end{vmatrix} = 9 + 12 = 21$$

$$\begin{pmatrix} 3 & -2 \\ 6 & 3 \end{pmatrix}^{-1} = \frac{1}{21} \begin{pmatrix} 3 & 2 \\ -6 & 3 \end{pmatrix}$$

Task 3.21

$$\begin{vmatrix} 2 & 1 & -1 \\ 0 & 2 & 1 \\ 5 & 2 & -3 \end{vmatrix} = -12 + 5 + 10 - 4 = -1$$

$$\begin{pmatrix} 2 & 1 & -1 \\ 0 & 2 & 1 \\ 5 & 2 & -3 \end{pmatrix}^{-1} =$$

$$-\begin{pmatrix} +[2 \cdot (-3) - (1 \cdot 2)] & -[0 \cdot (-3) - (1 \cdot 5)] & +[0 \cdot 2 - 2 \cdot 5] \\ -[1 \cdot (-3) - (-1 \cdot 2)] & +[2 \cdot (-3) - (-1 \cdot 5)] & -[2 \cdot 2 - 1 \cdot 5] \\ +[1 \cdot 1 - (-1 \cdot 2)] & -[2 \cdot 1 - (-1 \cdot 0)] & +[2 \cdot 2 - 1 \cdot 0] \end{pmatrix}^T$$

$$= -\begin{pmatrix} -8 & 5 & -10 \\ 1 & -1 & 1 \\ 3 & -2 & 4 \end{pmatrix}^T$$

$$= -\begin{pmatrix} -8 & 1 & 3 \\ 5 & -1 & -2 \\ -10 & 1 & 4 \end{pmatrix}$$

$$= \begin{pmatrix} 8 & -1 & -3 \\ -5 & 1 & 2 \\ 10 & -1 & -4 \end{pmatrix}$$

Task 3.22

$$2x_1 + x_2 - x_3 = 0$$
$$x_2 + x_3 = 0$$

We start with the first equation:

$$\text{I} - \text{II}: \quad 2x_1 - 2x_3 = 0$$
$$\boxed{x_1 = \lambda}: \quad 2\lambda - 2x_3 = 0$$
$$\boxed{x_3 = \lambda}$$

and use the parametric λ. Then we solve the second equation:

$$x_3 = \lambda \text{ in II}: \quad x_2 + \lambda = 0$$
$$\boxed{x_2 = -\lambda}.$$

and get the parametric solution or general solution $(x_1, x_2, x_3) = (\lambda, -\lambda, \lambda)$. The matrices are:

$$A = \begin{pmatrix} 2 & 1 & -1 \\ 0 & 1 & 1 \end{pmatrix} \rightarrow \text{Rank is 2}$$

$$(A \mid b) = \begin{pmatrix} 2 & 1 & -1 & 0 \\ 0 & 1 & 1 & 0 \end{pmatrix} \rightarrow \text{Rank is 2.}$$

We have 3 variables but only 2 equations. One of the variables has to be chosen (or treated as a parameter) and the solution for the 2 other variables will depend on this choice or the parameter. Because we have infinite different choices for this parameter (often called lambda λ), we have infinite solutions.

Task 3.23

$$2x_1 + x_2 = 0$$
$$x_1 - x_2 = 0$$

$$\text{I} + \text{II}: \quad 3x_1 = 0$$
$$x_1 = 0$$
$$x_1 = 0 \text{ in II}: \quad 0 - x_2 = 0$$
$$x_2 = 0$$

The matrices are:

$$A = \begin{pmatrix} 2 & 1 \\ 1 & -1 \end{pmatrix} \rightarrow \text{Rank is 2}$$

$$(A \mid b) = \begin{pmatrix} 2 & 1 & 0 \\ 1 & -1 & 0 \end{pmatrix} \rightarrow \text{Rank is 2.}$$

We have 2 equations and 2 variables with exactly 1 solution.

Task 3.24

$$x_1 + 2x_2 = 1$$
$$x_1 + 2x_2 = 2$$

$$\text{I} - \text{II}: \quad 0 = -1 \quad \text{\color{red}{ϟ}}$$

The equation set is inconsistent. The matrices are:

$$A = \begin{pmatrix} 1 & 2 \\ 1 & 2 \end{pmatrix} \rightarrow \text{Rank is 1}$$

$$(\,A \mid b\,) = \begin{pmatrix} 1 & 2 & \mid 1 \\ 1 & 2 & \mid 2 \end{pmatrix} \rightarrow \text{Rank is 2.}$$

Task 3.25

$$5x_1 + x_2 = 2$$
$$x_1 - 2x_2 = 7$$

Alternative 1:

$$2 \cdot \text{I} + \text{II}: \quad 11x_1 = 11$$
$$x_1 = 1$$
$$x_1 = 1 \text{ in II}: \quad 1 - 2x_2 = 7$$
$$x_2 = -3$$

The matrices are:

$$A = \begin{pmatrix} 5 & 1 \\ 1 & -2 \end{pmatrix} \rightarrow \text{Rank is 2}$$

$$(\,A \mid b\,) = \begin{pmatrix} 5 & 1 & \mid 2 \\ 1 & -2 & \mid 7 \end{pmatrix} \rightarrow \text{Rank is 2.}$$

We have 2 equations and 2 variables with exactly 1 solution.

Alternative 2:
We use Cramer's Rule now with the determinants:

$$\det(A) = \begin{vmatrix} 5 & 1 \\ 1 & -2 \end{vmatrix} = -10 - 1 = -11$$

$$\det(A_1) = \begin{vmatrix} 2 & 1 \\ 7 & -2 \end{vmatrix} = -4 - 7 = -11$$

$$\det(A_2) = \begin{vmatrix} 5 & 2 \\ 1 & 7 \end{vmatrix} = 35 - 2 = 33$$

which finally give:

$$x_1 = \frac{\det(A_1)}{\det(A)} = \frac{-11}{-11} = 1$$

$$x_1 = \frac{\det(A_2)}{\det(A)} = \frac{33}{-11} = -3$$

Task 3.26

$$2x_1 + x_2 - x_3 = -5$$
$$x_2 + x_3 = 1$$

Alternative 1:
We start with the first equation:

$$\text{I} + \text{II}: \quad 2x_1 + 2x_2 = -4$$
$$\boxed{x_1 = \lambda}: \quad 2\lambda + 2x_2 = -4$$
$$\boxed{x_2 = -2 - \lambda}$$

and then we solve the second equation:

$$x_2 = -2 - \lambda \text{ in II}: \quad -2 - \lambda + x_3 = 1$$
$$\boxed{x_3 = 3 + \lambda}.$$

Alternative 2:
2 equations with 3 unknowns. 1 is flexible. Choose $\boxed{x_3 = \lambda}$.

$$\text{I} - \text{II}: \quad 2x_1 - 2x_3 = -6$$
$$\boxed{x_1 = -3 + \lambda}$$
$$\text{II}: x_2 + x_3 = 1$$
$$\boxed{x_2 = 1 - \lambda}$$

The matrices are:

$$A = \begin{pmatrix} 2 & 1 & -1 \\ 0 & 1 & 1 \end{pmatrix} \rightarrow \text{Rank is 2}$$

$$(\,A \mid b\,) = \begin{pmatrix} 2 & 1 & -1 & \mid 5 \\ 0 & 1 & 1 & \mid 1 \end{pmatrix} \rightarrow \text{Rank is 2.}$$

More variables than equations. Thus, we have infinite solutions.

The number of solutions of linear equation sets can also be determined with the Rouché–Capelli theorem[5] summarized in Table 1.

---

[5] French mathematician: Eugène Rouché (1832—1910).
Italian mathematician: Alfredo Capelli (1855—1910).

Table 1. Overview: Solutions of linear equation systems (Rouché–Capelli theorem)

	homogeneity	
	$Ax = b$ m equations n variables	$Ax = 0$ (homogeneous system)
$Rk(A\|b) \neq Rk(A)$	system unsolvable **Task 3.24**	not possible for $b = 0$ homogeneous systems always solvable
$Rk(A\|b) = Rk(A)$ $m = n$	system solvable	
	unique solution **Task 3.25**	trivial solution $(x = 0)$ **Task 3.23**
$m < n$	non-unique solution **Task 3.26**	non-trivial solutions **Task 3.22**

Notes

Chapter 2: Metabolic modeling

Thomas Sauter, Marco Albrecht

Motivation

Metabolism is a mirror of many biological processes and extracellular metabolites can be accessed with relatively robust measurements. Moreover, basic analysis with a stoichiometric matrix (representing a metabolic network) can be adopted by many researchers because it is mainly based on linear algebra. However, well-trained experts are needed for the analysis of dynamic behavior, contextualization, and advanced research questions. Metabolism can also be analyzed on a large scale, *e.g.* with graph-theoretical approaches or in more detail with deep, mechanistic, and dynamical insights. In this chapter, we introduce **Stoichiometric Network Analysis (SNA)** which is an approach that is somewhere between detailed and large scale. SNA helps to exclude biologically unrealistic scenarios and, instead, gives us a variety of possible flux distributions. To thereof identify the biologically meaningful solutions, we need to constrain the solution space further, *e.g.* with measurements. We have two possible methods at hand. Either we impose known fluxes and maximize, *e.g.* the growth rate in a constrained model with **Flux Balance Analysis (FBA)**, or we impose measured fluxes and minimize residuals during numeric matrix manipulations with **Metabolic Flux Analysis (MFA)**. FBA also helps us to solve genome-wide problems, while MFA scrutinizes smaller models and allows the integration of isotope-labelled metabolites. Either way, the insight we obtain is crucial to understand the biology of metabolism.

Keywords

Metabolic modeling — Stoichiometric matrix — Flux Balance Analysis — Constraint-Based Modeling

Contact: thomas.sauter@uni.lu. **Licence**: CC BY-NC

Contents

1. Lecture summary

We learned about the essentials of graph theory and derived the first stoichiometric matrix (Chapter 1). Nodes and the interaction between them represent the topology of networks. Such networks can be written in matrix form. The introduction to linear algebra was of great importance to entering the field of metabolic modeling. Metabolic networks mainly describe enzyme-controlled fluxes of metabolites within a system. We can now study different methods for the modeling of metabolic networks except for dynamical mechanistic models (to be introduced in Chapter 3). Metabolic models can be built with different levels of detail and in various ways. We start with the following simple classification, in which the capital letters in the reactions are the **concentrations of substances** and the reaction arrows or edges define the **reactions**, which can symbolize enzymes, fluxes, and other transport systems. Principles of mechanism-based models and dynamic analysis will be addressed in Chapters 3 and 4 of this book.

Table 1. Bottom-up approaches:

interaction-based	graph theory methods
	(topology)
stoichiometry-based	stoichiometric network analysis
$A + B \rightleftharpoons C$	
mechanism-based	dynamic analysis
$A + B \underset{k_{-1}}{\overset{k_1}{\rightleftharpoons}} C + D$	

https://doi.org/10.11647/OBP.0291.02

1.1 ■ Stoichiometric Network Analysis
The stoichiometric matrix

For balancing, we need:

m... number of substances

r... number of reactions

n... stoichiometric coefficients

v_j... rate of reaction j

n_{ij}... stoichiometric coefficient (conversion) of metabolite i in reaction j

S_i... concentration of metabolite i (also: c_i)

These are elements of the following general equation that describes the change of substrate concentration S_i over time:

$$\frac{dS_i}{dt} = \dot{S}_i = \sum_{j=1}^{r} n_{ij} v_j \qquad i = 1,\ldots,m$$

$$= n_{i1} v_1 + n_{i2} v_2 + \ldots + n_{ir} v_r$$

The change of substrate concentrations is a consequence related to both the general stoichiometry of the reaction network—as indicated in the stoichiometric matrix N—and the given flux vector v. The respective matrix representation is:

$$\dot{S} = Nv \qquad \begin{aligned} S &= (S_1, S_2, \ldots, S_n) \\ v &= (v_1, v_2, \ldots, v_r) \end{aligned}$$

and the **stoichiometric matrix** is:

$$N = \{n_{ij}\} \qquad \begin{aligned} i &= 1 \ldots m \\ j &= 1 \ldots r \end{aligned}.$$

The columns of the stoichiometric matrix represent the reactions and the rows represent the metabolite concentrations.

For example, a reconstruction of a "genome-scale network" for *E. coli* encompasses 1136 unique metabolites and 2251 reactions. This results in a stoichiometric matrix N with 1136 rows and 2251 columns resulting in 2557136 entries. However, the matrix is usually sparse and only >25000 entries are actually non-zero [1]. The stoichiometric matrix:

- does not change over time (invariant)

- usually has many zero entries

- contains pseudo-reactions (uptake, excretion, transport, growth)

- is crucial for structural analysis with linear algebra.

Concentrations are usually written with brackets [], but in the context of this chapter we always assume concentrations when using a given molecule name in an equation, and thus write concentrations without brackets if not stated differently. *E.g.* we write G6P for the concentration of glucose-6-phosphate instead of [G6P]. Also note that some reactions can be reversible. In this case, we can only consider the net reaction rate $v_f - v_b$ (from left to right) to create the stoichiometric matrix. Thus, some fluxes can be negative, if the backward reaction v_b is larger than the forward reaction v_f. Alternatively, we can split a reversible reaction into two elementary reactions (see Example 1).

Example 1: Glucose to fructose

PGI reaction: $\quad G_6P \rightleftharpoons F_6P$
Two metabolites $m = 2$
Case 1: two elementary reactions (r=2):

PGI reaction: $\quad G_6P \underset{v_2}{\overset{v_1}{\rightleftharpoons}} F_6P \quad$ v=const.

$n_{11} = -1; \; n_{21} = 1; \; n_{12} = 1; \; n_{22} = -1$

$$N = \begin{pmatrix} -1 & 1 \\ 1 & -1 \end{pmatrix}$$

$$\Rightarrow \quad \dot{S}_1 = \dot{G_6P} = n_{11}v_1 + n_{12}v_2 = -v_1 + v_2$$
$$\dot{S}_2 = \dot{F_6P} = n_{21}v_1 + n_{22}v_2 = +v_1 - v_2$$

Case 2: one net reaction (r=1):

PGI: $\quad G_6P \overset{v}{\rightleftharpoons} F_6P \quad$ v=const.

$n_{11} = -1; \; n_{21} = 1$

$$N = \begin{pmatrix} -1 \\ 1 \end{pmatrix}$$

$$\Rightarrow \quad \dot{S}_1 = \dot{G_6P} = n_{11}v = -v$$
$$\dot{S}_2 = \dot{F_6P} = n_{21}v = +v$$

Example 2: Oxohydrogen reaction

Reaction: $\quad 2H_2 + O_2 \overset{v_1}{\rightleftharpoons} 2H_2O$
$m = 3; \; r = 1; \; n_{11} = -2; \; n_{21} = -1; \; n_{31} = +2$

$$\Rightarrow \quad \begin{aligned} \dot{S}_1 &= \dot{H_2} &= (-2v_f + 2v_b) &= -2v_1 \\ \dot{S}_2 &= \dot{O_2} &= (-v_f + v_b) &= -v_1 \\ \dot{S}_3 &= \dot{H_2O} &= (+2v_f - 2v_b) &= +2v_1 \end{aligned}$$

The stoichiometric matrix is:

$$
\overbrace{\underbrace{\begin{array}{c} v_1 \end{array}}}^{\text{reactions}}
$$

$$
N = \left.\begin{pmatrix} -2 \\ -1 \\ 2 \end{pmatrix}\begin{matrix} H_2 \\ O_2 \\ H_2O \end{matrix}\right\}\text{metabolites}
$$

with the final representation in matrix and equation set form:

$$
\begin{pmatrix} \dot{H_2} \\ \dot{O_2} \\ \dot{H_2O} \end{pmatrix} = \begin{pmatrix} -2 \\ -1 \\ 2 \end{pmatrix}(v_1) \quad\Leftrightarrow\quad \begin{array}{ll} \dot{H_2} & = -2v_1 \\ \dot{O_2} & = -v_1 \\ \dot{H_2O} & = +2v_1 \end{array}
$$

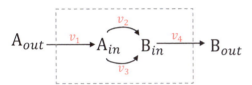

Figure 1. A simple metabolic system. The grey dotted line represents the system boundary. The A Molecule is transported from **out**side to **in**side and is then converted to Molecule B via two possible reactions. Finally, Molecule B is leaving the system.

Example 3: Simple metabolic network

As shown in Figure 1, there are two possible ways to produce Metabolite B, either via v_2 or v_3. All network edges can be weighted so that we do not have an incidence matrix, but a stoichiometric matrix:

$$
N = \begin{bmatrix} & v_1 & v_2 & v_3 & v_4 & \\ & 1 & -1 & -1 & 0 & \end{bmatrix}\begin{matrix} A_{in} \\ & 0 & 1 & 1 & -1 & \end{matrix}\quad
$$

$$
N = \begin{bmatrix} 1 & -1 & -1 & 0 \\ 0 & 1 & 1 & -1 \\ -1 & 0 & 0 & 0 \\ 0 & 0 & 0 & 1 \end{bmatrix}\begin{matrix} A_{in} \\ B_{in} \\ A_{out} \\ B_{out} \end{matrix}
$$

where the reactions determine the columns and the metabolite concentration determines the row entries. External metabolites are not balanced in this approach because the steady state of external metabolites (in our example A_{out} and B_{out}) cannot be assumed. Consequently, we can reduce the system to:

$$
N = \begin{bmatrix} 1 & -1 & -1 & 0 \\ 0 & 1 & 1 & -1 \end{bmatrix}\begin{matrix} A \\ B \end{matrix}
$$

Example 4: Enzyme kinetics

Balancing: $E + S \underset{v_{-1}}{\overset{v_1}{\rightleftharpoons}} ES \overset{v_2}{\to} E+P$

Resulting in:

$$
\underset{(4x1)}{\begin{pmatrix} \dot{S} \\ E \\ ES \\ P \end{pmatrix}} = \underset{(4x3)}{\begin{pmatrix} -1 & 1 & 0 \\ -1 & 1 & 1 \\ 1 & -1 & -1 \\ 0 & 0 & 1 \end{pmatrix}}\underset{(3x1)}{\begin{pmatrix} v_1 \\ v_{-1} \\ v_2 \end{pmatrix}}
$$

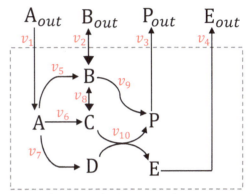

Figure 2. Another metabolic network/system with two inputs and three outputs. The transport of Molecule B is bi-directional. The system is controlled by the transport of the external concentration of A and B and produces Product P but can also produce B and E.

Example 5: Larger metabolic network

A slightly larger metabolic network is shown in Figure 2 with the reversible reactions:

$$
rev = \{R2, R8\}
$$

and irreversible reactions:

$$
irrev = \{R1, R3, R4, R5, R6, R7, R9, R10\}
$$

Note that the forward reactions ($v_2 : B_{out} \to B$ and $v_8 : B \to C$) were considered in order to create the stoichiometric matrix N:

$$
N = \begin{bmatrix}
v_1 & v_2 & v_3 & v_4 & v_5 & v_6 & v_7 & v_8 & v_9 & v_{10} \\
1 & 0 & 0 & 0 & -1 & -1 & -1 & 0 & 0 & 0 \\
0 & 1 & 0 & 0 & 1 & 0 & 0 & -1 & -1 & 0 \\
0 & 0 & 0 & 0 & 0 & 1 & 0 & 1 & 0 & -1 \\
0 & 0 & 0 & 0 & 0 & 0 & 1 & 0 & 0 & -1 \\
0 & 0 & 0 & -1 & 0 & 0 & 0 & 0 & 0 & 1 \\
0 & 0 & -1 & 0 & 0 & 0 & 0 & 0 & 1 & 1
\end{bmatrix}\begin{matrix} A \\ B \\ C \\ D \\ E \\ P \end{matrix}
$$

Example 6: Branched metabolic network

$$\overset{v_1}{\rightleftharpoons} S_1 \overset{v_2}{\rightleftharpoons} 2S_2 \overset{v_4}{\rightleftharpoons}$$

$$\downarrow\!\!\downarrow v_3$$

$$S_3$$

What does the stoichiometric matrix N and the balance equation look like?
The stoichiometric matrix is:

$$\overbrace{\begin{matrix} v_1 & v_2 & v_3 & v_4 \end{matrix}}^{reactions}$$
$$N = \left.\begin{pmatrix} 1 & -1 & -1 & 0 \\ 0 & 2 & 0 & -2 \\ 0 & 0 & 1 & 0 \end{pmatrix}\right\}\begin{matrix} S_1 \\ S_2 \\ S_3 \end{matrix} \text{ metabolites}$$

The balance equation of the system in the matrix form is:

$$\underset{(3x1)}{\begin{pmatrix} \dot{S_1} \\ S_2 \\ S_3 \end{pmatrix}} = \underset{(4x4)}{\begin{pmatrix} 1 & -1 & -1 & 0 \\ 0 & 2 & 0 & -2 \\ 0 & 0 & 1 & 0 \end{pmatrix}} \underset{(4x1)}{\begin{pmatrix} v_1 \\ v_2 \\ v_3 \\ v_4 \end{pmatrix}}$$

and in the equation set form is:

$$\dot{S_1} = v_1 - v_2 - v_3$$
$$\dot{S_2} = 2v_2 - 2v_4$$
$$\dot{S_3} = v_3.$$

System in steady state
Usually we would like to look closer at our system and see what happens if:

- fluxes are constant and do not change over time

- metabolite concentrations are constant and do not change over time ($\dot{S} = 0$).

Therefore, we look at the steady-state solution:

$$Nv = 0$$

The steady-state solution is available and appropriate as soon as the states of the system no longer change and thus time dynamics are not taking place.
See Example 7 with Figure 3. Here, we have the situation as described in the motivation: we have many non-trivial solutions but we do not yet know which solution is biologically meaningful. Therefore, we need to constrain the system further by integrating experimental data. In the practical work, one rate might be measured, *e.g.* $v_1 = 1$ and then the other fluxes (v_2, v_3) can be calculated accordingly.

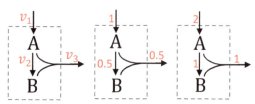

Figure 3. Flux distributions of Example 7. The balance of input flux v_1 and output flux v_3 requires a detour via Metabolite B with flux v_2.

Example 7: Flux in steady state

The system in Figure 3 can be written as a system of elementary reactions:

$$\emptyset \overset{v_1}{\rightleftharpoons} A$$

$$A \overset{v_2}{\rightleftharpoons} B$$

$$A + B \overset{v_3}{\rightleftharpoons} \emptyset$$

and the related equation set:

$$\dot{A} = v_1 - v_2 - v_3 = 0 \tag{1}$$
$$\dot{B} = v_2 - v_3 = 0 \tag{2}$$

which can be written in the matrix form:

$$\underbrace{\begin{bmatrix} 1 & -1 & -1 \\ 0 & 1 & -1 \end{bmatrix}}_{N} \underbrace{\begin{bmatrix} v_1 \\ v_2 \\ v_3 \end{bmatrix}}_{v} = 0 \qquad \mathrm{rk}(N) = 2 < 3 = r$$

Because the rank of N is smaller than the number of reactions, we have to deal with an infinite number of non-trivial solutions.
Solution: Assume: $v_1 = a$.
Beginning with Equation 1:

$$a - v_2 - v_3 = 0$$
$$v_2 = a - v_3$$

which is then put in Equation 2:

$$a - v_3 - v_3 = 0$$
$$a = 2v_3$$
$$v_3 = \frac{a}{2}$$

which in turn gives the solution if it is put into the previous equation $\Rightarrow v_2 = a - \frac{a}{2} = \frac{a}{2}$. The solution is:

$$\begin{bmatrix} v_1 \\ v_2 \\ v_3 \end{bmatrix} = a \begin{bmatrix} 1 \\ 0.5 \\ 0.5 \end{bmatrix} \qquad ; a \in \mathbb{R}$$

Elementary flux modes (EFM)

If we cannot determine a unique solution of the flux network due to too many unknown rates (see also rank of the stoichiometric matrix), then we can study elementary flux modes and the kernel matrix, both of which allow us to describe all possible solutions. Some applications thereof are:

- testing a set of enzymes for the production of a desired product

- detecting non-redundant pathways

- analyzing effects of enzyme deficiency

- identifying drug targets.

Flux modes describe possible pathways from one metabolite to another if the system is in steady state. Elementary flux modes (EFM) are a set of non-unique and linear-independent basis vectors v_k, which can be summarized in the kernel matrix K. The kernel matrix also fulfils the steady-state relationship:

$$NK = 0$$

Example 3 delivers for the molecules A and B the following stoichiometric matrix:

$$N = \underbrace{\begin{bmatrix} 1 & -1 & -1 & 0 \\ 0 & 1 & 1 & -1 \end{bmatrix}}_{r-rk(N)=4-2=2} \rightarrow K = \begin{bmatrix} 1 & 2 \\ 1 & 1 \\ 0 & 1 \\ 1 & 2 \end{bmatrix}$$

The kernel matrix K has as many columns as the stoichiometric matrix N has linear-dependent columns (number of columns/reactions r minus the rank of N). The determination of the kernel matrix is shown in Case Box 1. The EFMs in the kernel matrix are non-unique and can thus consist of all vectors which:

- fulfil the steady-state condition

- are linear-independent of each other.

Non-unique means that the kernel matrix can have many different solutions which fulfil the steady-state relationship $NK = 0$. Often, we are just interested in getting one of these possible solutions. Now, you could say, you find more flux modes than the indicated basis vectors (EFM) in the kernel matrix K. Well, the flux modes v can be reconstructed by the non-unique basis vectors (EFM) in the kernel matrix v_k:

$$v_k : \underbrace{\begin{bmatrix} 1 \\ 1 \\ 0 \\ 1 \end{bmatrix} \& \begin{bmatrix} 2 \\ 1 \\ 1 \\ 2 \end{bmatrix}}_{\text{independent}} \quad v : \underbrace{\begin{bmatrix} 1 \\ 0 \\ 1 \\ 1 \end{bmatrix} ; \begin{bmatrix} -1 \\ -1 \\ 0 \\ -1 \end{bmatrix} ; \begin{bmatrix} 10 \\ 5 \\ 5 \\ 10 \end{bmatrix} ; \begin{bmatrix} -3 \\ -1 \\ -2 \\ -3 \end{bmatrix} ; \dots}_{\text{dependent}}$$

with column labels: I, II for v_k; II−I, −I, 5II, I−2II for v.

In our example, another possible kernel matrix would be:

$$K = \begin{bmatrix} 1 & 1 \\ 1 & 0 \\ 0 & 1 \\ 1 & 1 \end{bmatrix}$$

The kernel matrix can provide information about **blocked reactions**. Blocked reactions are pathways without a metabolic flux. Rows in the kernel matrix which have only zero entries, *e.g.*:

$$K = \begin{bmatrix} \dots & \dots & \dots \\ 0 & \cdot 0 \cdot & 0 \\ \dots & \dots & \dots \end{bmatrix} \rightarrow \text{blocked.}$$

indicate blocked reactions and two examples can be found in Figure 4a and 4b.

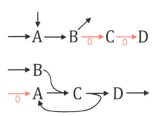

(a) Blocked reactions by dead-end or lack of input.

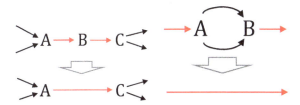

(b) Coupled reactions reduced (left) and null-space analysis via the kernel matrix (right).

The kernel matrix also refers to **coupled reactions**. Coupled reactions are reactions which always appear together in the same ratio. Their rows only differ—if at all—in a scalar factor:

$$K = \begin{bmatrix} 1 & 0 \\ 0 & -1 \\ 1 & 1 \\ 1 & 0 \end{bmatrix}$$

Blocked reactions could be removed and coupled reactions could be simplified, which would reduce the number of columns r in the stoichiometric matrix so that we reach another network representation with the property $r - \text{rk} N = 0$. Remember that, depending on the rank of the stoichiometric matrix, we have two possibilities for a solvable homogeneous system according to Table 2.

Table 2. Overview: Resolvability of linear equation systems for metabolic models (Rouché–Capelli theorem).

N has m metabolites r reactions	homogeneity	
	$Nv = b$ (inhomogeneous system)	$Nv = 0$ (homogeneous system)
$\mathrm{rk}(N\|b) \neq \mathrm{rk}(N)$	system unsolvable	not possible for $b = 0$ homogeneous systems always solvable
$\mathrm{rk}(N\|b) = \mathrm{rk}(N)$	system solvable	
$m = r$	unique solution	trivial solution $(v = 0)$
$m < r$	non-unique solution	non-trivial solutions

We either have the trivial solution or the non-trivial solution:

Trivial solution: thermodynamic equilibrium with all fluxes equal to zero $v = 0$.

Non-trivial solution: under-determined system with $\mathrm{rk}(N)$

Case 1: Calculate the kernel matrix

A kernel matrix only contains a linear-independent subset of flux modes, not all. Thus it is a non-unique matrix. Let's start with the example:

$$N = \begin{bmatrix} 1 & -1 & -1 & 0 \\ 0 & 1 & 1 & -1 \end{bmatrix}$$

Nullity = Number of necessary basis vectors:

$$r - \mathrm{rk}(N) = 4 - 2 = 2$$

We want to know how many of the reactions r are not fully determined by the stoichiometric matrix with rank $\mathrm{rk}(N)$.

1. Basis vector:

$$N k_1 = 0$$

$$\begin{bmatrix} 1 & -1 & -1 & 0 \\ 0 & 1 & 1 & -1 \end{bmatrix} \begin{bmatrix} k_{11} \\ k_{12} \\ k_{13} \\ k_{14} \end{bmatrix} = 0$$

which is as an equation set:

$$\Rightarrow \quad k_{11} - k_{12} - k_{13} = 0 \qquad \text{I}$$
$$k_{12} + k_{13} - k_{14} = 0 \qquad \text{II}$$

We have 4 unknowns and 2 equations. So, we can set 2 to arbitrary values, *e.g.*:

$$k_{11} = 1; \quad k_{12} = 1$$

whereby the value 1 is the most frequently used number to simplify follow up calculations. The simplification and cleanliness of calculation reduces the risk of errors and is thus a smarter choice. We get, for the remaining unknowns, the values:

$$\Rightarrow \quad k_{13} = 0$$
$$\Rightarrow \quad k_{14} = 1$$

and the basis vector:

$$\Rightarrow \quad k_1 = \begin{pmatrix} 1 \\ 1 \\ 0 \\ 1 \end{pmatrix}$$

2. Basis vector:

$$N k_2 = 0$$

$$\begin{bmatrix} 1 & -1 & -1 & 0 \\ 0 & 1 & 1 & -1 \end{bmatrix} \begin{bmatrix} k_{21} \\ k_{22} \\ k_{23} \\ k_{24} \end{bmatrix} = 0$$

which is as an equation set:

$$\Rightarrow \quad k_{21} - k_{22} - k_{23} = 0 \qquad \text{I}$$
$$k_{22} + k_{23} - k_{24} = 0 \qquad \text{II}$$

We have 4 unknowns and 2 equations. So, we can set two unknowns to arbitrary values, *e.g.*:

$$k_{21} = 2; \quad k_{22} = 1$$

With this choice, we get for the remaining unknowns:

$$\Rightarrow \quad k_{23} = 1$$
$$\Rightarrow \quad k_{24} = 2$$

and a second basis vector:

$$\Rightarrow \quad k_2 = \begin{pmatrix} 2 \\ 1 \\ 1 \\ 2 \end{pmatrix}$$

Building a kernel matrix:

$$\Rightarrow K = (k_1, k_2) = \left(\begin{pmatrix} 1 \\ 1 \\ 0 \\ 1 \end{pmatrix} \begin{pmatrix} 2 \\ 1 \\ 1 \\ 2 \end{pmatrix} \right) = \begin{pmatrix} 1 & 2 \\ 1 & 1 \\ 0 & 1 \\ 1 & 2 \end{pmatrix}$$

Conservation relations for metabolites

If the number of balanced metabolites exceeds the rank of the stoichiometric matrix, we can identify conservation relations such as:

$$[NADH] + [NAD] = \text{const}$$
$$[ATP] + [ADP] = \text{const}$$

This means that some of the metabolites are dependent on other metabolites and do not act independently. Such conservation relations can be identified by calculating:

$$N^T Y = 0$$

A linear-independent subset of the conservation relations build basis vectors in the matrix Y. The construction is shown in Case Box 2.

Case 2: Calculate the Y matrix

A given stoichiometric matrix is:

$$N = \begin{bmatrix} -1 \\ -1 \\ 1 \\ 1 \end{bmatrix}$$

Nullity = Number of necessary basis vectors:

$$m - \text{rk}(N) = 4 - 1 = 3$$

We want to know which metabolites can be seen as dependent on other metabolites in our system. In our case, we have only 1 reaction but 4 metabolites. Thus, we have only 1 independent metabolite. The remaining 3 metabolites depend on the selected independent metabolite.

1. Basis vector:

$$N^T y_1 = 0$$

$$\begin{bmatrix} -1 & -1 & 1 & 1 \end{bmatrix} \begin{bmatrix} y_{11} \\ y_{12} \\ y_{13} \\ y_{14} \end{bmatrix} = 0$$

is the same as the equation:

$$\Rightarrow \quad -y_{11} - y_{12} + y_{13} + y_{14} = 0$$

We have 4 unknowns and 1 equation. So, we can set 3 to arbitrary values, *e.g.*:

$$y_{11} = 1; \quad y_{12} = -1; \quad y_{13} = 0$$

With this choice, we get for the remaining variable:

$$\Rightarrow \quad y_{14} = 0$$

and the non-unique elementary flux mode:

$$\Rightarrow \quad y_1 = \begin{pmatrix} 1 \\ -1 \\ 0 \\ 0 \end{pmatrix}$$

2. Basis vector:

$$N^T y_2 = 0$$

$$\begin{bmatrix} -1 & -1 & 1 & 1 \end{bmatrix} \begin{bmatrix} y_{21} \\ y_{22} \\ y_{23} \\ y_{24} \end{bmatrix} = 0$$

is the same as the equation:

$$\Rightarrow \quad -y_{21} - y_{22} + y_{23} + y_{24} = 0$$

We have 4 unknowns and 1 equation. So, we can set 3 to arbitrary values, *e.g.*:

$$y_{21} = 1; \quad y_{22} = 0; \quad y_{23} = 1$$

With this choice, we get:

$$\Rightarrow \quad y_{24} = 0$$

and the non-unique elementary flux mode:

$$\Rightarrow \quad y_2 = \begin{pmatrix} 1 \\ 0 \\ 1 \\ 0 \end{pmatrix}$$

3. Basis vector:

$$N^T y_3 = 0$$

$$\begin{bmatrix} -1 & -1 & 1 & 1 \end{bmatrix} \begin{bmatrix} y_{31} \\ y_{32} \\ y_{33} \\ y_{34} \end{bmatrix} = 0$$

is the same as the equation:

$$\Rightarrow \quad -y_{31} - y_{32} + y_{33} + y_{34} = 0$$

We have 4 unknowns and 1 equation. So, we can set 3 to arbitrary values, *e.g.*:

$$y_{31} = 0; \quad y_{32} = 1; \quad y_{33} = 0$$

With this choice, we get:

$$\Rightarrow \quad y_{34} = 0$$

and the non-unique elementary flux mode:

$$\Rightarrow \quad \mathbf{y}_3 = \begin{pmatrix} 0 \\ 1 \\ 0 \\ 1 \end{pmatrix}$$

Building a conservation relation matrix:

$$\Rightarrow \mathbf{Y} = \left(\begin{pmatrix} 1 \\ -1 \\ 0 \\ 0 \end{pmatrix} \begin{pmatrix} 1 \\ 0 \\ 1 \\ 0 \end{pmatrix} \begin{pmatrix} 0 \\ 1 \\ 0 \\ 1 \end{pmatrix} \right) = \begin{pmatrix} 1 & 1 & 0 \\ -1 & 0 & 1 \\ 0 & 1 & 0 \\ 0 & 0 & 1 \end{pmatrix}$$

Each non-unique but independent conservation relation in matrix $\mathbf{Y}$ refers to one linear-dependent row of $\mathbf{N}$. Two applications can be studied in the Examples 8 and 9.

Example 8: Conservation relations

From the simple hypergraph:
$A + B \rightleftharpoons C + D$
we can create the stoichiometric matrix $\mathbf{N}$:

$$\mathbf{N} = \begin{bmatrix} -1 \\ -1 \\ 1 \\ 1 \end{bmatrix}$$

We have 3 linear-independent basis vectors ($m - \text{rk}(\mathbf{N})$) because the stoichiometric matrix $\mathbf{N}$ has 4 species (rows) and one reaction (rank 1).
In order to find the conservation relations, we calculate the conservation relation matrix using the transposition of $\mathbf{N}$, denoted $\mathbf{N}^T$. The approach to calculating $\mathbf{Y}$ was shown in a previous section.

$$\mathbf{N}^T \mathbf{Y} = \mathbf{0} = \begin{bmatrix} -1 & -1 & 1 & 1 \end{bmatrix} \mathbf{Y} = 0$$

$$\underset{(4-1)=3}{\longrightarrow} \mathbf{Y} = \begin{matrix} & \overset{y_1}{} & \overset{y_2}{} & \overset{y_3}{} & \\ \begin{bmatrix} 1 & 1 & 0 \\ -1 & 0 & 1 \\ 0 & 1 & 0 \\ 0 & 0 & 1 \end{bmatrix} & \begin{matrix} A \\ B \\ C \\ D \end{matrix} \end{matrix}$$

The constraint matrix does not only constrain the network behavior, it also tells us whether a dataset is of good quality. The constraint matrix multiplied with the metabolite concentrations $c(t)$ is always the same:

$$\mathbf{Y}^T c(t=0) = \text{const} = \mathbf{Y}^T c(t)$$

even if compared to the initial concentration $c(0)$. This is useful when we want to perform a consistency check, scrutinized in detail in Example 9.

Example 9: Consistency check

We have a system with 4 metabolites and 2 reactions:
$D \rightarrow A$ and $2A + B \rightarrow C$.
The initial concentrations are:
$c_A(t=0) = c_B(0) = c_C(0) = c_D(0) = 2$.
Two hard-working experimental biologists measure (independently of each other) the concentrations at the same time point, t:
    Exp1: $c_A(t) = c_B(t) = c_D(t) = 1$ and $c_C(t) = 3$.
    Exp2: $c_A(t) = c_B(t) = 3$ and $c_C(t) = c_D(t) = 1$.
Can they be right?

1) Stoichiometric matrix:

$$\mathbf{N} = \begin{bmatrix} 1 & -2 \\ 0 & -1 \\ 0 & 1 \\ -1 & 0 \end{bmatrix}$$

If we apply the calculation of conservation relations:

$$\mathbf{N}^T \mathbf{Y} = \mathbf{0} = \begin{bmatrix} 1 & 0 & 0 & -1 \\ -2 & -1 & 1 & 0 \end{bmatrix} \mathbf{Y} = \begin{bmatrix} 0 \\ 0 \end{bmatrix}$$

You have two possibilities. You can either set $\mathbf{Y} = 0$, which is the trivial solution no one is interested in, or you find non-zero values for this matrix. After Gaussian elimination, we obtain the non-trivial solution:

$$\mathbf{N}^T \mathbf{Y} = \begin{bmatrix} 1 & 0 & 0 & -1 \\ -2 & -1 & 1 & 0 \end{bmatrix} \begin{bmatrix} y_1 \\ y_2 = \lambda_2 \\ y_3 \\ y_4 = \lambda_1 \end{bmatrix} = \begin{bmatrix} 0 \\ 0 \end{bmatrix}$$

We can set two unknowns to the arbitrary values $y_4 = \lambda_1$, $y_2 = \lambda_2$. Consequently, we obtain for the remaining unknowns:

$$y_1 = \lambda_1$$
$$y_3 = 2\lambda_1 + \lambda_2$$

$$\mathbf{y} = \begin{bmatrix} \lambda_1 \\ \lambda_2 \\ 2\lambda_1 + \lambda_2 \\ \lambda_1 \end{bmatrix} = \lambda_1 \overset{y^1}{\begin{bmatrix} 1 \\ 0 \\ 2 \\ 1 \end{bmatrix}} + \lambda_2 \overset{y^2}{\begin{bmatrix} 0 \\ 1 \\ 1 \\ 0 \end{bmatrix}} \overset{\lambda_{1,2}=1}{\longrightarrow} \mathbf{Y} = \begin{bmatrix} 1 & 0 \\ 0 & 1 \\ 2 & 1 \\ 1 & 0 \end{bmatrix}$$

After we obtain the $\mathbf{Y}$ matrix, we can perform the

consistency check with:

$$Y^T c(t) = \text{const}$$

$$
\begin{bmatrix} 1 & 0 & 2 & 1 \\ 0 & 1 & 1 & 0 \end{bmatrix}
\begin{array}{c} {\scriptstyle t=0} \ {\scriptstyle Exp1} \ {\scriptstyle Exp2} \end{array}
\begin{bmatrix} 2 & 1 & 3 \\ 2 & 1 & 3 \\ 2 & 3 & 1 \\ 2 & 1 & 1 \end{bmatrix}
=
\begin{array}{c} {\scriptstyle t=0} \ {\scriptstyle Exp1} \ {\scriptstyle Exp2} \end{array}
\begin{bmatrix} 8 & 8 & 6 \\ 4 & 4 & 4 \end{bmatrix}
$$

or written differently:

$$t_0 : y^1 = 2+0+2\cdot2+2 = 8; \quad y^2 = 0+2+2+0 = 4$$
$$\text{Exp1} : y^1 = 1+0+2\cdot3+1 = 8; \quad y^2 = 0+1+3+0 = 4$$
$$\text{Exp2} : y^1 = 3+0+2\cdot1+1 = 6; \quad y^2 = 0+3+1+0 = 4$$

This indicates that Experimental Biologist 2 had a bad day and made a mistake.

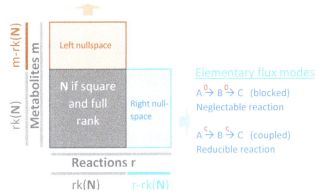

Conservation relations

A= ς - B (e.g. ATP= const − ADP)
Metabolite A depends on B

Figure 5. Nullspace scheme. If the column space is larger than the rank, we can obtain elementary flux modes. If the row space is larger than the rank, we can relate some metabolites to the metabolites being considered independent.

Nullspace to identify conserved metabolites or coupled and blocked reactions:
The nullspace represents linear-dependent vectors that fall into the origin of the independent vectors, as shown in Figure 5. An excess of metabolites can be described as dependent on the core metabolites and then the left nullspace disappears. An excess of reactions can contain either blocked reactions, which can be ignored, or coupled reactions, which can be pooled. In both cases, the column space can be reduced after the identification of the elementary flux modes.
We learned that a linear-independent subset of elementary flux modes is in the right nullspace NK, and that a

linear-independent subset of the conservation relations is in the left nullspace:

$$Y^T N = 0^T \Leftrightarrow N^T Y = 0.$$

Both matrices contain basis vectors, from which infinite many vectors can be generated. The matrix N is either multiplied by the kernel K from the right or by the matrix Y from the left, as shown in Figure 6. This makes sense because we know that the stoichiometric matrix N contains the information of the reactions/fluxes in the columns and the metabolites in the rows. During matrix multiplication, matrices from the left conserve the row space of N representing metabolites (conservation relations) and matrices from the right conserve the column space of N, representing reactions (stationary fluxes). Remember that the number of columns of the left matrix must equal the number of the rows of the right matrix:

with α, β, and γ representing the number of columns and rows.

Example 10: Demonstration of equivalence

Show that:

$$N^T Y = 0 \Leftrightarrow Y^T N = 0^T$$

is true for Example 9:
The stoichiometric matrix N:

$$N = \begin{bmatrix} -1 \\ -1 \\ 1 \\ 1 \end{bmatrix}$$

and the conservation relation matrix Y are:

$$\begin{bmatrix} 1 & 1 & 0 \\ -1 & 0 & 1 \\ 0 & 1 & 0 \\ 0 & 0 & 1 \end{bmatrix}$$

Case 1: $N^T Y = 0$

$$\begin{bmatrix} -1 & -1 & 1 & 1 \end{bmatrix} \begin{bmatrix} 1 & 1 & 0 \\ -1 & 0 & 1 \\ 0 & 1 & 0 \\ 0 & 0 & 1 \end{bmatrix}$$
$$= \begin{bmatrix} -1+1+0+0 & -1-0+1+0 & 0-1+0+1 \end{bmatrix}$$
$$= \begin{bmatrix} 0 & 0 & 0 \end{bmatrix}$$

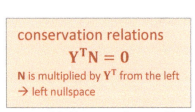 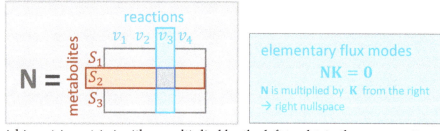

Figure 6. Nullspace scheme: The stoichiometric matrix is either multiplied by the left to obtain the conservation of metabolites or multiplied by the right to obtain conserved reactions.

Case 2: $Y^T N = 0^T$

$$
\begin{bmatrix} 1 & -1 & 0 & 0 \\ 1 & 0 & 1 & 0 \\ 0 & 1 & 0 & 1 \end{bmatrix} \begin{bmatrix} -1 \\ -1 \\ 1 \\ 1 \end{bmatrix}
$$

$$
= \begin{bmatrix} (1\cdot-1)+(-1\cdot-1)+(0\cdot1)+(0\cdot1) \\ (1\cdot-1)+(0\cdot-1)+(1\cdot1)+(0\cdot1) \\ (0\cdot-1)+(1\cdot-1)+(0\cdot1)+(1\cdot1) \end{bmatrix}
$$

$$
= \begin{bmatrix} -1+1+0+0 \\ -1+0+1+0 \\ 0-1+0+1 \end{bmatrix}
$$

$$
= \begin{bmatrix} 0 \\ 0 \\ 0 \end{bmatrix}
$$

1.2 ■ Integration of experimental data

The balance equation $Nv = 0$ is frequently underdetermined and rates have to be measured. A smart idea to separate the balance equation into a **m**easured part and an **u**nknown part:

$$Nv = 0$$

$$N \begin{pmatrix} v_u \\ v_m \end{pmatrix} = 0$$

$$N_u v_u + N_m v_m = 0$$

$$N_u v_u = -N_m v_m$$

$$v_u = N_u^{-1}(-N_m v_m)$$

to obtain an inhomogeneous linear equation set.

As you can see, we are interested in v_u and need the inverse of the stoichiometric matrix. The inverse N_u^{-1} is often not available, and the measured matrix entries are hardly exact in reality due to measurement noise, which leads to inconsistencies.

The basic integration of data can be studied in a well-posed Example 11 with Figure 7, and more generally in Example 12 with four typical scenarios.

Figure 7. Reduced network of glycolysis from glucose to phosphoenolpyruvate with measured and unknown rates.

Example 11: Glucose to phosphoenolpyruvic acid

For this small metabolic network and model (Figure 7), we assume that all components outside the system boundary can be neglected. The system boundary can be freely chosen and might, in this case, be the double-lipid layer of the cell membrane. We are interested in the steady-state concentration of glucose-6-phosphate (G_6P), dihydroxyacetonephosphate (DHAP) and phosphoenolpyruvate (PEP). The balance equation for those metabolites in steady state is:

$$\begin{pmatrix} \dot{G_6P} \\ \dot{DHAP} \\ \dot{PEP} \end{pmatrix} = 0$$

$$\Leftrightarrow \begin{pmatrix} 1 & -1 & -1 & 0 & 0 \\ 0 & 0 & 2 & -1 & 0 \\ 0 & 0 & 0 & 1 & -1 \end{pmatrix} \begin{pmatrix} v_1 \\ v_2 \\ v_3 \\ v_4 \\ v_5 \end{pmatrix} = 0$$

If we insert the measured values from Figure 7, we obtain:

$$\Leftrightarrow \begin{pmatrix} 1 & -1 & -1 & 0 & 0 \\ 0 & 0 & 2 & -1 & 0 \\ 0 & 0 & 0 & 1 & -1 \end{pmatrix} \begin{pmatrix} 1 \\ v_2 \\ v_3 \\ v_4 \\ 1.4 \end{pmatrix} = 0$$

Now we can calculate:

$$\begin{pmatrix} 1 & -1 & -1 & 0 & 0 \\ 0 & 0 & 2 & -1 & 0 \\ 0 & 0 & 0 & 1 & -1 \end{pmatrix} \begin{pmatrix} 1 \\ v_2 \\ v_3 \\ v_4 \\ 1.4 \end{pmatrix} = 0.$$

$$\Leftrightarrow \begin{pmatrix} -1 & -1 & 0 \\ 0 & 2 & -1 \\ 0 & 0 & 1 \end{pmatrix} \begin{pmatrix} v_2 \\ v_3 \\ v_4 \end{pmatrix} = -\begin{pmatrix} 1 & 0 \\ 0 & 0 \\ 0 & -1 \end{pmatrix} \begin{pmatrix} 1 \\ 1.4 \end{pmatrix}$$

$$\Leftrightarrow \begin{pmatrix} -1 & -1 & 0 \\ 0 & 2 & -1 \\ 0 & 0 & 1 \end{pmatrix} \begin{pmatrix} v_2 \\ v_3 \\ v_4 \end{pmatrix} = \begin{pmatrix} -1 \\ 0 \\ 1.4 \end{pmatrix}$$

We now multiply each side by the inverse from the left to obtain the identity matrix:

$$N^{-1}N = I$$

$$\begin{pmatrix} 1 & 0 & 0 \\ 0 & 1 & 0 \\ 0 & 0 & 1 \end{pmatrix} \begin{pmatrix} v_2 \\ v_3 \\ v_4 \end{pmatrix} = \begin{pmatrix} -1 & -1 & 0 \\ 0 & 2 & -1 \\ 0 & 0 & 1 \end{pmatrix}^{-1} \begin{pmatrix} -1 \\ 0 \\ 1.4 \end{pmatrix}$$

$$\begin{pmatrix} v_2 \\ v_3 \\ v_4 \end{pmatrix} = \begin{pmatrix} -1 & -\frac{1}{2} & -\frac{1}{2} \\ 0 & \frac{1}{2} & \frac{1}{2} \\ 0 & 0 & 1 \end{pmatrix} \begin{pmatrix} -1 \\ 0 \\ 1.4 \end{pmatrix}$$

$$\begin{pmatrix} v_2 \\ v_3 \\ v_4 \end{pmatrix} = \begin{pmatrix} 0.3 \\ 0.7 \\ 1.4 \end{pmatrix}$$

and now we know that the fluxes in Figure 7 are $v_2 = 0.3$, $v_3 = 0.7$, and $v_4 = 1.4$.

Measurement scenarios

Stoichiometric Network Analysis delivers physiological snapshots and is good for testing "if-then" scenarios. It also helps us to see how measured flux rates influence other rates via sensitivity analysis. However, the measured rates are often not sufficient, and loops and alternative pathways might cause problems.

The dissection of our stoichiometric matrix into measured and unknown parts reveals the problem of calculating with an inverse matrix:

$$v_u = N_u^{-1}(-N_m v_m)$$

As the inverse N_u^{-1} is often not available and measurements are never perfectly correct, we use an approximated matrix. The Moore–Penrose pseudo-inverse[1][2][3] $N_u^{\#}$ is available for all matrices. Any mathematicians or physicists among you might enjoy a review of this method [2]; all others might just use the *MATLAB* command B = pinv(A). Moreover, we use the kernel matrix K_u with related arbitrary vector a so that $K_u a = v$. The final equation might be:

$$v_u = -N_u^{\#} N_m v_m + K_u a$$

whereby we have 4 cases, which depend on the number of unknown reactions x and the number of metabolites m. The 4 cases are:

- **determined:** $\mathrm{rk}(N_u) = x$. All rates can be determined.

- **underdetermined:** $\mathrm{rk}(N_u) < x$. Not all rates can be determined.

- **not redundant:** $\mathrm{rk}(N_u) = m$.

- **redundant:** $\mathrm{rk}(N_u) < m$. Inconsistencies likely.

and are presented in the following matrix as an example:

$$N = \begin{bmatrix} v_1 & v_2 & v_3 & v_4 \\ 1 & -1 & -1 & 0 \\ 0 & 1 & 1 & -1 \end{bmatrix} \begin{matrix} A \\ B \end{matrix}$$

Example 12: Principal measurement scenarios

Here are four different cases for stoichiometric network analysis. Red fluxes are measured.
a) The first case describes a determined, non-redundant system. We know that the input of the systems needs to be equal to its output, thus $v_1 = v_4 = 2$. As $v_2 = 0$, and $v_2 + v_3 = v_1 = v_4 = 2$, we can easily calculate $v_3 = 2$.

[1] American mathematician Eliakim Hastings Moore (1862—1932).
[2] Swedish geodesist Arne Bjerhammar (1917—2011).
[3] English mathematical physicist Roger Penrose (1931—today).

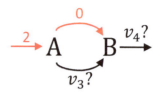

b) The second case describes an underdetermined, non-redundant system. We know that the input of the systems needs to be equal to its output, thus $v_1 = v_4 = 2$. The fluxes v_2 and v_3 do not correspond to a single solution but the sum of their fluxes needs to be equal to $v_2 + v_3 = v_1 = v_4 = 2$.

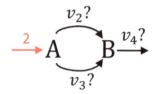

c) The third case describes a determined and redundant system, in which v_1 and v_4 are inconsistent. The input $v_1 = 2$ does not correspond to the output $v_4 = 3$.

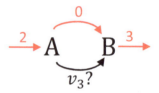

d) The last case describes an underdetermined and redundant system, in which v_1 and v_4 are inconsistent.

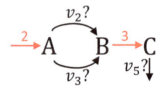

Case 1: determined, not redundant (Ex. 12a)

-given:

$$v_m = \begin{bmatrix} v_1 \\ v_2 \end{bmatrix} = \begin{bmatrix} 2 \\ 0 \end{bmatrix}; \quad x = 2$$

$$N_m = \begin{bmatrix} v_1 & v_2 \\ 1 & -1 \\ 0 & 1 \end{bmatrix}; \quad N_u = \begin{bmatrix} v_3 & v_4 \\ -1 & 0 \\ 1 & -1 \end{bmatrix}$$

-check: $\mathrm{rk}(N_u) = x = m = 2 \rightarrow$ determined, not redundant

-result:

$$K_u = \begin{bmatrix} 0 \\ 0 \end{bmatrix}; \quad N_u^\# = N_u^{-1} = \begin{bmatrix} -1 & 0 \\ -1 & -1 \end{bmatrix};$$

$$v_u = \begin{bmatrix} v_3 \\ v_4 \end{bmatrix} = -N_u^\# N_m v_m + K_u a = \begin{bmatrix} 2 \\ 2 \end{bmatrix}$$

Case 2: underdetermined, not redundant (Ex. 12b)

-given:

$$v_m = \begin{bmatrix} v_1 \end{bmatrix} = \begin{bmatrix} 2 \end{bmatrix}; \quad x = 3$$

$$N_m = \begin{matrix} v_1 \\ \begin{bmatrix} 1 \\ 0 \end{bmatrix} \end{matrix}; \quad N_u = \begin{matrix} v_2 \quad v_3 \quad v_4 \\ \begin{bmatrix} -1 & -1 & 0 \\ 1 & 1 & -1 \end{bmatrix} \end{matrix}$$

-check: $\mathrm{rk}(N_u) = m = 2 < 3 = x \to$ underdetermined, not redundant

-result:

$$K_u = \begin{bmatrix} 1 \\ -1 \\ 0 \end{bmatrix}; \quad N_u^\# = \begin{bmatrix} -0.5 & 0 \\ -0.5 & 0 \\ -1 & -1 \end{bmatrix};$$

$$v_u = \begin{bmatrix} v_2 \\ v_3 \\ v_4 \end{bmatrix} = -N_u^\# N_m v_m + K_u a = \begin{bmatrix} 1 \\ 1 \\ 2 \end{bmatrix} + \begin{bmatrix} 1 \\ -1 \\ 0 \end{bmatrix} a$$

Case 3: determined, redundant (Ex. 12c)

-given:

$$v_m = \begin{bmatrix} v_1 \\ v_2 \\ v_4 \end{bmatrix} = \begin{bmatrix} 2 \\ 0 \\ 3 \end{bmatrix}; \quad x = 1$$

$$N_m = \begin{matrix} v_1 \quad v_2 \quad v_4 \\ \begin{bmatrix} 1 & -1 & 0 \\ 0 & 1 & -1 \end{bmatrix} \end{matrix}; \quad N_u = \begin{matrix} v_3 \\ \begin{bmatrix} -1 \\ 1 \end{bmatrix} \end{matrix}$$

-check: $\mathrm{rk}(N_u) = 1 = x < 2 = m \to$ determined, redundant

-result:

$$K_u = \begin{bmatrix} 0 \end{bmatrix}; \quad N_u^\# = \begin{bmatrix} -0.5 & 0.5 \end{bmatrix};$$

Inconsistent and redundant scenarios might be forced to consistency with $\hat{v}_1 = \hat{v}_4 = 2.5$ (SSR, com-

promise between 2 and 3):

$$v_u = \begin{bmatrix} v_3 \end{bmatrix} = -N_u^\# N_m \hat{v}_m + K_u a$$

$$= -\begin{bmatrix} -0.5 & 0.5 \end{bmatrix} \begin{bmatrix} 1 & -1 & 0 \\ 0 & 1 & -1 \end{bmatrix} \begin{bmatrix} 2.5 \\ 0 \\ 2.5 \end{bmatrix} + \begin{bmatrix} 0 \end{bmatrix} a$$

$$= 2.5$$

With weighted estimator processes, one could thereby consider the variance (measurement error).

Case 4: underdetermined, redundant (Ex.12d)

-given:

$$v_m = \begin{bmatrix} v_1 \\ v_4 \end{bmatrix} = \begin{bmatrix} 2 \\ 3 \end{bmatrix}; \quad x = 2$$

$$N_m = \begin{matrix} v_1 \quad v_4 \\ \begin{bmatrix} 1 & 0 \\ 0 & -1 \end{bmatrix} \end{matrix}; \quad N_u = \begin{matrix} v_2 \quad v_3 \\ \begin{bmatrix} -1 & -1 \\ 1 & 1 \end{bmatrix} \end{matrix}$$

-check: $\mathrm{rk}(N_u) = 1 < m = x = 2 \to$ underdetermined, redundant

-result:

$$K_u = \begin{bmatrix} 1 \\ -1 \end{bmatrix}; \quad N_u^\# = \begin{bmatrix} -0.25 & 0.25 \\ -0.25 & 0.25 \end{bmatrix} \to \text{get no rate};$$

Inconsistent, redundant scenarios might be forced to consistency with $\hat{v}_1 = \hat{v}_4 = 2.5$ (SSR):

$$v_u = \begin{bmatrix} v_2 \\ v_3 \end{bmatrix} = -N_u^\# N_m \hat{v}_m + K_u a$$

$$= -\begin{bmatrix} -0.25 & 0.25 \\ -0.25 & 0.25 \end{bmatrix} \begin{bmatrix} 1 & 0 \\ 0 & -1 \end{bmatrix} \begin{bmatrix} 2.5 \\ 2.5 \end{bmatrix} + \begin{bmatrix} 1 \\ -1 \end{bmatrix} a$$

$$= \begin{bmatrix} 1.25 \\ 1.25 \end{bmatrix} + \begin{bmatrix} 1 \\ -1 \end{bmatrix} a$$

An underdetermined and redundant system can also have rates, which we can calculate. An example is rate v_5.

1.3 ■ Constraint-Based Modeling

Flux Balance Analysis (FBA) [3] is a widely used method within Stoichiometric Network Analysis to determine possible flux distributions. It can be conducted with a more general Constraint-Based Modeling approach (CBM), thereby allowing the integration of more information beyond the steady-state assumption. Possible applications are:

- Prediction of phenotypes
- Prediction of mutant behavior.

We will now step-by-step introduce FBA and CBM. In the beginning, the solution space can be considered unconstrained, meaning that any reaction can carry any flux. Then, we apply constraints, such as the mass balance constraints imposed by the stoichiometric matrix S.
YouTube: FBA

The basic idea: start from all possible fluxes and incorporate further constraints to limit network behavior to a smaller and thus more informative solution space.

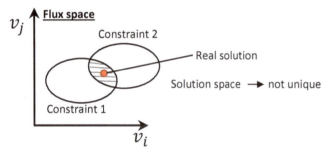

The lower v_{min} and upper v_{max} bounds of a reaction are thereby also taken into account.

Main applied constraints

- Mass conservation: ($\mathbf{N}\mathbf{v} = \mathbf{0}$)
The material that is entering via the influx does not get lost or accumulate (steady-state assumption). The outgoing fluxes need to balance the incoming fluxes for each balanced metabolite. However, the fluxes are often given in mole per time. Metabolites might be split during biochemical reactions and thus the respective molar fluxes multiply accordingly. But the mass remains the same.

$$\dot{A} = v_1 - v_2 - v_3 \overset{!}{=} 0$$
$$\Rightarrow v_2 + v_3 = v_1$$

e.g. $v_1 = 1$:
Case 1: $v_2 = 0 \rightarrow v_3 = 1 \rightarrow$ mark point: $(v_2 = 0; v_3 = 1)$

Case 2: $v_3 = 0 \rightarrow v_2 = 1 \rightarrow$ mark point: $(v_2 = 1; v_3 = 0)$
Case 3: $v_3 = \frac{1}{2} \rightarrow v_2 = \frac{1}{2} \rightarrow$ mark point: $(v_2 = \frac{1}{2}; v_3 = \frac{1}{2})$

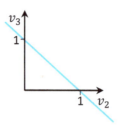

- Reaction reversibility:
Irreversible: $v_j \geq 0$
Reversible: $-\inf < v_j < \inf$

e.g. $v_2, v_3 \geq 0$ (irreversible)

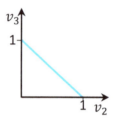

- Boundary conditions:
imposed by experimental setup or measurements, e.g., stationary continuous cultivation:

$$\mu = D \qquad (e.g. = 0.1\mathrm{h}^{-1})$$
$$v_j = \hat{v}_j = \mathrm{const.}$$

- Enzyme capacities:
e.g. $v_j \leq v_{j,max}$ (maximal catalysis rate):

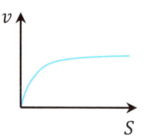

- Thermodynamic constraints: based on chemical potentials:

$$v_{j,min} \leq v_j \leq v_{j,max}$$

e.g. with:

$$0.5 \leq v_1 \leq 1$$

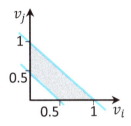

The constraint-based model:
We end up with a constraint-based model for which we can use an optimization algorithm—such as linear programming—to find the best possible flux combination to maximize (or minimize) a given objective function $F = C^T v$. The coefficient matrix C thereby defines which flux(es) shall be optimized. This optimization is constrained by the system in the steady state itself $Nv = 0$ but also by all incorporated equality and inequality constraints:

$$\underset{v_{\text{aim}}}{\text{maximize}} \quad C^T v = c_1 v_1 + c_2 v_2 + \ldots + c_r v_r$$

$$\text{subject to} \qquad Nv = 0$$
$$0 \leq v_j \qquad \text{; some } j$$
$$v_j = \hat{v}_j = \text{const.} \quad \text{; some } j$$
$$v_{j,\text{min}} \leq v_j \leq v_{j,\text{max}} \quad \text{; some } j$$

We obtain a system of linear equality and inequality constraints. This also defines a cone in the v-space, as shown in Figure 8.

Example 13: Apply the constraints and optimize

We have the simple system:

with:

- mass constraint: $v_1 = v_2 + v_3$
- reaction irreversible: $v_2 \geq 0, v_3 \geq 0$
- enzyme capacity: $v_1 \leq 1$

Solutions space? Unique solution?

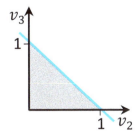

The idea behind this is that biological systems are evolutionarily optimized to maximize a certain behavior such as:

- maximal growth yield (microorganism)
- maximal growth rate (microorganism)
- maximal energy production (mitochondria)
- maximal yield of product (genetic manipulation by human).

We optimize the objective function F with:

$$F = C^T \cdot v \to \text{max}$$

$$\text{e.g.} \quad F = \begin{bmatrix} 0 & 1 & 0 \end{bmatrix} \begin{bmatrix} v_1 \\ v_2 \\ v_3 \end{bmatrix} \to \text{max}$$

$$F = v_2 \to \text{max}$$

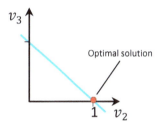

The solution is obtained by <u>linear</u> optimization, which is fast and reliable. We use software for it, although we have the problem that the algorithms usually compute only 1 optimal solution. Other equally optimal solutions might exist and need to be determined by corresponding algorithms (like flux variability analysis or random sampling). We could now also include an enzyme capacity *e.g.* $v_2 \leq 0.6$:

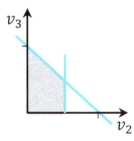

Then, the optimization does not result in a single solution. Instead all values $0 \leq v_3 \leq 0.4$ are possible.
What would happen if we impose $v_3 = 0.5$?

$$\Rightarrow v_1 = v_2 + v_3 = 0.6 + 0.5 = 1.1 \leq 1 \quad \text{⚡}$$

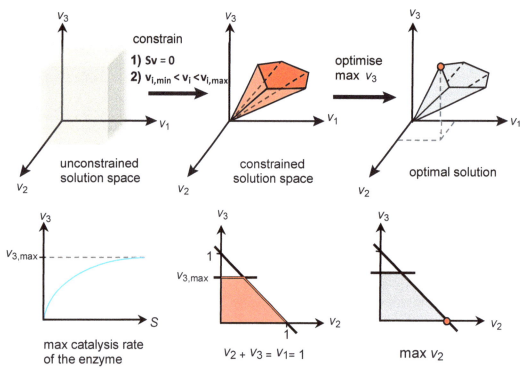

Figure 8. FBA and constraint-based modeling.
Top: Stoichiometric Network Analysis might leave a large solution space. Due to the mass balance constraints imposed by the stoichiometric matrix S and the upper and lower bounds of a reaction, one obtains a constrained solution space (red). This space is used by optimization algorithms to find one optimal solution for biomass (for example).

Bottom: One constraint captures the maximum turnover velocity of an enzyme $v_{3,max} < 1$, which is combined with another constraint $v_1 = 1$ from measuring this rate. We know then that the sum of the v_2 and v_3 fluxes must be 1 as well (dark lines). The red area shows the constrained solution space for an alternative v_1 constraint: $0 \leq v_1 \leq 1$. After maximizing v_2, it becomes clear that the enzyme responsible for v_3 has to be inactive in order to maximize v_2. Adapted from [3]. Copyright © 1969, Nature Publishing Group.

Example 14: FBA example

Given a simple network:

$$\xrightarrow{v_1} S \xrightarrow{v_2}$$
$$\downarrow v_3$$

with mass conservation $\mathbf{N}\mathbf{v} = \mathbf{0}$ leading to:

$$\dot{S} = v_1 - v_2 - v_3 \overset{!}{=} 0$$
$$v_1 = v_2 + v_3.$$

We assume that all reactions are irreversible and the respective rates are thus positive:

$$v_2, v_3 \geq 0$$

We can introduce boundary conditions, such as that the influx is fixed:

$$v_1 = 1$$

We can further assume that some fluxes depend on enzymes with limited capacity $v_3 \leq v_{3,max}$ and assume that one flux such as through the biomass reaction has to be maximized:

$$\max F = \mathbf{C}^T \mathbf{v} = \begin{bmatrix} 0 & 1 & 0 \end{bmatrix} \begin{bmatrix} v_1 \\ v_2 \\ v_3 \end{bmatrix} = v_2$$

The optimal solution is $v_2 = 1$.

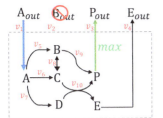 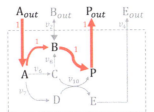 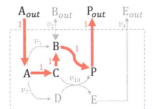 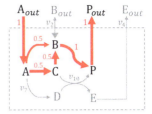

Figure 9. Flux solutions for Example 15. From left to right: The original pathway model has been modified to maximize P_{out} with fixed A_{out} and no B_{out}. We obtain two extreme linear-independent solutions: the first solution with direct flux from A to B and the second solution indirect via C. The last scheme represents infinite solutions with different ratios of the two extreme pathways from A to B.

Example 15: Opt. flux balance for P_{ext} without B_{ext}

We use the metabolic network shown in Example 5 and Figure 2. We search for the optimal flux distribution for blocked B_{ext} and maximized P_{ext} as shown in Figure 9 with **objective function $F = C^T v$**:

$$\text{maximize } C^T v = (0,0,1,0,0,0,0,0,0,0)v = v_3$$

With constraints:

- $Nv = 0$, according to Example 5.

- Set fluxes $v_1 = 1$ (uptake) and $v_2 = 0$ (no uptake) according to the available measurement information.

- Lower bounds $v_{min} = (1,0,0,0,0,0,0,-\inf,0,0)$
  Upper bounds $v_{max} = (1,0,\inf,\ldots,\inf)$.

The resulting infinite solutions (linear combinations from the first two solutions) are illustrated in Figure 9.
Always check uniqueness.

Optimization after knockout:
Another optimization example is shown in Figure 10. The central metabolism across different related organisms is especially well-conserved while pathways with less evolutionary pressure might be less optimized. One can also assume that organisms optimize their program after knockout experiments, as shown in Figure 11 and studied *e.g.* by Segre *et. al.* [4].

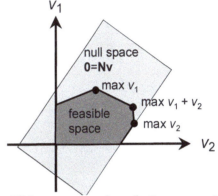

Figure 10. FBA constrains the solution space. If we only consider fluxes v_1 and v_2, we can either optimize for v_1 or v_2 alone, or we can optimize for the sum of the fluxes of $v_1 + v_2$.

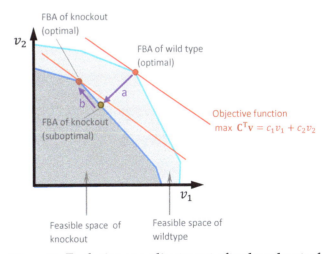

Figure 11. Evolutionary adjustment after knockout of a gene and its respective reactions. After a knockout, the feasible space is reduced and a suboptimal solution might be present. With time, the evolutionary optimization step b results in a new optimal flux distribution in the reduced solution space. Figure guided by [4].

1.4 ■ Problems & discussion

False positive and false negative prediction of gene/reaction knockouts

False **positive** prediction (the organism does not live in reality after the knockout although the model states that it should):

- Kinetics / regulation not considered in the model

- Neglected side-effects, *e.g.* toxic intermediates.

False **negative** prediction (the organism does live in reality although the model states that it should not):

- Proof for incorrect network structure → Quality control & iterative model refinement possible.

Limited applicability

- Steady-state assumption → no dynamics considered, *e.g.* accumulation of metabolites

- Focus on mass transfer → barely applicable to cellular information processing.

Objective function

- Strong dependence of results on choice of objective function

- Use of 'natural' objective functions such as growth:

  - Not applicable to all organisms (*e.g.* cells in multicellular organisms → cancer)

  - Not applicable under all conditions (*e.g.* after perturbation of an organism)

- Alternative/ conflicting numerical approaches for optimization

- Alternative approach for fluxes: principles of flux minimization ($\sim$ effort for establishing a network).

Alternative optima

- Linear programming problem: finding a solution can be guaranteed

- Unique value of the objective function ('growth')
- Existence of infinitely many optimal solutions with optimal value of objective possible.

- Without incorporating further constraints often poor performance in predicting flux distributions.

1.5 ■ Metabolic Flux Analysis

Metabolic flux analysis (MFA), together with ^{13}C-labelled isotope experiments, improves the characterization of fluxes and is called isotope-based flux analysis. Different experiments using labelled isotopes are summarized in Figure 12 together with the two possibilities of stoichiometric network analysis without isotopes. We do not go into detail but you are encouraged to study Figure 13 and the referenced paper.

Metabolic flux analysis gives us deeper insights into metabolic processes. See also the following reviews [5, 6]. The models remain small. See also Figure 14 for a comparison with the previously described Flux Balance Analysis (FBA). FBA returns a larger solution space, which is more helpful for large metabolic networks. However, these solutions might contain different flux distributions that can all maximize the assumed cellular objective, *i.e.* growth or biomass production [5]. The different principles of MFA and FBA are reflected by two major optimization strategies in computational biology shown in Figure 15. Either one imposes the measurements and either tries to reduce the sum of squared residuals, or one defines boundaries in which the function is forced to stay.

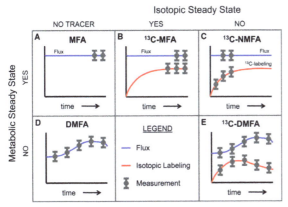

Figure 12. Isotope experiments for dynamic and steady-state MFA. The blue line represents metabolites and not fluxes. Source [5]. Copyright © 2015, Oxford University Press.

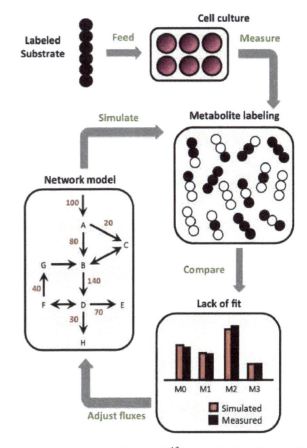

Figure 13. Isotope tracing and ^{13}C-metabolic flux analysis. "In simple metabolic networks, each pathway produces a unique labeling pattern in the final product, and the resulting mass isotopomer distribution provides a direct measure of relative flux in the network. Mass isotopomers are molecules with the same chemical formula but different molecular weights due to varying incorporation of heavy isotopes. They are denoted M0, M1, M2, etc., in order of increasing weight. In complex networks, a computational model is applied to determine fluxes by minimizing the lack of fit between simulated and measured labeling patterns at multiple pathway nodes. The flux parameters in the model are iteratively adjusted until the optimization converges". Direct quotation: [7]. Licence: CC BY-NC-SA 3.0.

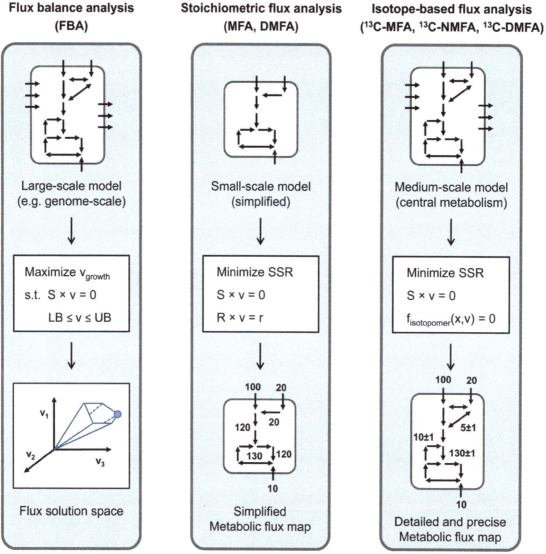

Figure 14. Overview on methods. Each of these methods uses mathematical optimization to approach the measured data. S x v = 0 corresponds here with $\boldsymbol{Nv} = \boldsymbol{0}$ and R x v = r corresponds here with $\boldsymbol{N_u v_u} = -\boldsymbol{N_m v_m}$. FBA: fluxes are balanced until they remain between defined lower (LB) and upper (UB) boundaries. MFA: numeric adjustment to minimize sum of square residuals. DMFA: dynamic metabolic flux analysis. SSR: sum of squared errors of prediction, either without isotopes (Stoichiometric Network Analysis) or with isotopes (isotope-based flux analysis). Source: [5]. Copyright © 2015, Oxford University Press.

Measurement-based fit (MFA)

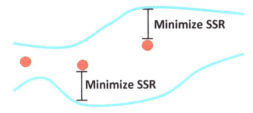

Constraint-based fit (FBA)

Figure 15. Conceptual illustration of fitting a function to measurement data vs. constraining it: Functions can be fitted in at least two ways. The most frequent is the measurement-based fit. Mean values (red dots) are imposed and a function (blue line) shall come close enough to the function in order to fit, usually by minimizing the sum of squared residuals (SSR). The second possibility is the imposing of constraints. The function has to stay within the bounds. The latter can also be based on measurements, e.g. by using mean +/ standard deviation as the maximum and minimum values.

References

[1] Jeffrey D Orth, Tom M Conrad, Jessica Na, Joshua A Lerman, Hojung Nam, Adam M Feist, and Bernhard Ø Palsson. A comprehensive genome-scale reconstruction of Escherichia coli metabolism—2011. *Molecular Systems Biology*, 7(1):535, 2011. https://doi.org/10.1038/msb.2011.65.

[2] João Carlos Alves Barata and Mahir Saleh Hussein. The moore–penrose pseudoinverse: A tutorial review of the theory. *Brazilian Journal of Physics*, 42(1-2):146–165, 2012. https://doi.org/10.1007/s13538-011-0052-z.

[3] Jeffrey D Orth, Ines Thiele, and Bernhard Ø Palsson. What is flux balance analysis? *Nature Biotechnology*, 28(3):245, 2010. https://doi.org/10.1038/nbt.1614.

[4] Daniel Segre, Dennis Vitkup, and George M Church. Analysis of optimality in natural and perturbed metabolic networks. *Proceedings of the National Academy of Sciences*, 99(23):15112–15117, 2002. https://doi.org/10.1073/pnas.232349399.

[5] Maciek R Antoniewicz. Methods and advances in metabolic flux analysis: a mini-review. *Journal of Industrial Microbiology & Biotechnology*, 42(3):317–325, 2015. https://doi.org/10.1007/s10295-015-1585-x.

[6] Maciek R Antoniewicz. A guide to 13 C metabolic flux analysis for the cancer biologist. *Experimental & Molecular Medicine*, 50(4):19, 2018. https://doi.org/10.1038/s12276-018-0060-y.

[7] Casey Scott Duckwall, Taylor Athanasaw Murphy, and Jamey Dale Young. Mapping cancer cell metabolism with 13C flux analysis: recent progress and future challenges. *Journal of Carcinogenesis*, 12, 2013. https://doi.org/10.4103/1477-3163.115422.

2. Exercises

■ **Stoichiometric matrix N**

a) Formulate N for the following biochemical system:

$$v_1 : E + S \rightleftharpoons ES$$
$$v_2 : ES \rightarrow E + P$$

b) Formulate N for the following biochemical system:

$$v_1 : E + S \rightleftharpoons ES$$
$$v_2 : ES \rightarrow E + P$$
$$v_3 : E + I \rightleftharpoons EI$$
$$v_4 : ES + I \rightleftharpoons ESI$$
$$v_5 : EI + S \rightleftharpoons ESI$$

c) Formulate N for the following biochemical reaction chain:

$$A \overset{v_1}{\rightleftharpoons} B \overset{v_2}{\rightleftharpoons} C \overset{v_3}{\rightleftharpoons} D \overset{v_4}{\rightleftharpoons} E \overset{v_5}{\rightleftharpoons} A$$

d) Extract the biochemical reaction of metabolites S_1 to S_5 from the following stoichiometric matrix N:

$$N = \begin{pmatrix} 1 & -1 & 0 & 0 \\ 0 & -2 & 0 & 0 \\ 0 & 1 & -1 & 0 \\ 0 & 3 & -1 & 0 \\ 0 & 0 & 2 & -1 \end{pmatrix}$$

■ **Flux Balance Analysis (FBA)**

The following biochemical equation system is given:

$$v_1 : \varnothing \rightarrow A$$
$$v_2 : \varnothing \rightarrow B$$
$$v_3 : A \rightarrow C$$
$$v_4 : B \rightarrow C$$
$$v_5 : B \rightarrow D$$
$$v_6 : C \rightarrow \varnothing$$
$$v_7 : D \rightarrow \varnothing$$

Under the steady-state condition, the following rates were measured: $v_1 = 1$, $v_2 = 1$, and $v_7 = 0.5$. Calculate the remaining rates v_3 to v_6, applying the methodology of Flux Balance Analysis (via N and by dividing the balance equation into known and unknown parts)! To make your life easier:

$$\begin{pmatrix} -1 & 0 & 0 & 0 \\ 0 & -1 & -1 & 0 \\ 1 & 1 & 0 & -1 \\ 0 & 0 & 1 & 0 \end{pmatrix}^{-1} = \begin{pmatrix} -1 & 0 & 0 & 0 \\ 0 & -1 & 0 & -1 \\ 0 & 0 & 0 & 1 \\ -1 & -1 & -1 & -1 \end{pmatrix}$$

■ **The kernel matrix K**

a) Calculate the kernel matrix K of the following stoichiometric matrix $N = \begin{pmatrix} 1 & 1 & -2. \end{pmatrix}$

b) Determine dead ends and unbranched reactions pathways, given the following kernel matrix K:

$$K = \begin{pmatrix} -1 & 1 & 0 & 0 \\ 0 & 0 & -1 & 2 \\ -1 & 1 & 0 & 0 \\ 0 & 0 & 0 & 0 \\ 1 & 2 & -1 & 0 \\ 0 & 0 & -1 & 2 \\ -1 & 1 & 0 & 0 \\ -2 & 0 & 0 & 1 \end{pmatrix}$$

■ **Constraint-Based Modeling of an example network**

An example network is composed of substrates (S_1, S_2), intracellular metabolites (A, B, C), and the biomass (X).

- The following reactions v_1 to v_6 take place:

$$v_1 : S_1 \rightarrow A$$
$$v_2 : S_2 \rightarrow B$$
$$v_3 : C \rightarrow X$$
$$v_4 : A \rightarrow C$$
$$v_5 : B \rightarrow C$$
$$v_6 : A \rightleftharpoons B$$

- Reactions v_1, v_2, v_3, v_4, v_5 are irreversible.

- For v_6, a maximal enzyme capacity is given: $v_6 \leqslant v_{6,max}$.

- Due to the experimental setting, v_3 is fixes to $v_3 = D$.

Questions:

a) Draw the biochemical network based on v_1 to v_6 and the irreversibility information.

b) Formulate the balance equations for the intracellular metabolites A, B, and C in steady state.

c) Replace v_3 by D in the resulting equations. Keep v_1 and v_6 as variables and solve for v_2, v_4, and v_5.

d) Apply the irreversibility information of v_1 to v_5 and extract therefore (several) inequality constraints for the remaining variables v_1 and v_6.

e) Mark the possible solution space using a v_6 over v_1 plot using the obtained inequality constraints and the enzyme capacity $v_{6,max}$. Distinguish thereby two cases:
  1) $v_{6,max} \geq D$
  2) $v_{6,max} < D$

f) Give the (unique) solution for the case $v_6 \geq D$ while optimizing for $v_6 \rightarrow max$ (Remember that when not otherwise indicated, the maximization of a reversible reaction implies the maximization of the production of the metabolites on the right side of the equation).

Notes

3. Solutions

Do not betray yourself!

◼ The stoichiometric matrix N

Task a)   Formulate N for the following biochemical system:

$$v_1 : E + S \rightleftharpoons ES$$
$$v_2 : ES \rightarrow E + P$$

The solution is:

$$N = \begin{array}{c} \\ \begin{array}{cc} v_1 & v_2 \end{array} \\ \begin{bmatrix} -1 & 1 \\ -1 & 0 \\ 1 & -1 \\ 0 & 1 \end{bmatrix} \begin{array}{c} E \\ S \\ ES \\ P \end{array} \end{array}$$

Task b)   Formulate N for the following biochemical system:

$$v_1 : E + S \rightleftharpoons ES$$
$$v_2 : ES \rightarrow E + P$$
$$v_3 : E + I \rightleftharpoons EI$$
$$v_4 : ES + I \rightleftharpoons ESI$$
$$v_5 : EI + S \rightleftharpoons ESI$$

The solution is:

$$N = \begin{array}{c} \\ \begin{array}{ccccc} v_1 & v_2 & v_3 & v_4 & v_5 \end{array} \\ \begin{bmatrix} -1 & 1 & -1 & 0 & 0 \\ -1 & 0 & 0 & 0 & -1 \\ 1 & -1 & 0 & -1 & 0 \\ 0 & 1 & 0 & 0 & 0 \\ 0 & 0 & -1 & -1 & 0 \\ 0 & 0 & 1 & 0 & -1 \\ 0 & 0 & 0 & 1 & 1 \end{bmatrix} \begin{array}{c} E \\ S \\ ES \\ P \\ I \\ EI \\ ESI \end{array} \end{array}$$

Task c)   Formulate N for the following biochemical reaction chain:

$$A \rightleftharpoons B \rightleftharpoons C \rightleftharpoons D \rightleftharpoons E \rightleftharpoons A$$

The solution is:

$$N = \begin{array}{c} \\ \begin{array}{ccccc} v_1 & v_2 & v_3 & v_4 & v_5 \end{array} \\ \begin{bmatrix} -1 & 0 & 0 & 0 & 1 \\ 1 & -1 & 0 & 0 & 0 \\ 0 & 1 & -1 & 0 & 0 \\ 0 & 0 & 1 & -1 & 0 \\ 0 & 0 & 0 & 1 & -1 \end{bmatrix} \begin{array}{c} A \\ B \\ C \\ D \\ E \end{array} \end{array}$$

Task d)   Extract the biochemical reaction of metabolites S_1 to S_5 from the following stoichiometric matrix N:

$$N = \begin{array}{c} \\ \begin{array}{cccc} v_1 & v_2 & v_3 & v_4 \end{array} \\ \begin{bmatrix} 1 & -1 & 0 & 0 \\ 0 & -2 & 0 & 0 \\ 0 & 1 & -1 & 0 \\ 0 & 3 & -1 & 0 \\ 0 & 0 & 2 & -1 \end{bmatrix} \begin{array}{c} S_1 \\ S_2 \\ S_3 \\ S_4 \\ S_5 \end{array} \end{array}$$

The solution is:

$$v_1 : \varnothing \rightarrow S_1$$
$$v_2 : S_1 + 2S_2 \rightarrow S_3 + 3S_4$$
$$v_3 : S_3 + S_4 \rightarrow 2S_5$$
$$v_4 : S_5 \rightarrow$$

◼ Flux Balance Analysis (FBA)

The following biochemical equation system is given:

$$v_1 : \varnothing \rightarrow A$$
$$v_2 : \varnothing \rightarrow B$$
$$v_3 : A \rightarrow C$$
$$v_4 : B \rightarrow C$$
$$v_5 : B \rightarrow D$$
$$v_6 : C \rightarrow \varnothing$$
$$v_7 : D \rightarrow \varnothing$$

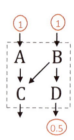

Under the steady-state condition, the following rates were measured: $v_1 = 1$, $v_2 = 1$, and $v_7 = 0.5$. Calculate the remaining rates v_3 to v_6, applying the methodology of flux balance analysis (via N and by dividing the balance equation into known and unknown parts)! To make your life easier:

$$\begin{pmatrix} -1 & 0 & 0 & 0 \\ 0 & -1 & -1 & 0 \\ 1 & 1 & 0 & -1 \\ 0 & 0 & 1 & 0 \end{pmatrix}^{-1} = \begin{pmatrix} -1 & 0 & 0 & 0 \\ 0 & -1 & 0 & -1 \\ 0 & 0 & 0 & 1 \\ -1 & -1 & -1 & -1 \end{pmatrix}$$

The solution is:
From the biochemical equation system we get:

$$N = \begin{array}{c} \\ \begin{array}{ccccccc} v_1 & v_2 & v_3 & v_4 & v_5 & v_6 & v_7 \end{array} \\ \begin{pmatrix} 1 & 0 & -1 & 0 & 0 & 0 & 0 \\ 0 & 1 & 0 & -1 & -1 & 0 & 0 \\ 0 & 0 & 1 & 1 & 0 & -1 & 0 \\ 0 & 0 & 0 & 0 & 1 & 0 & -1 \end{pmatrix} \begin{array}{c} A \\ B \\ C \\ D \end{array} \end{array}$$

Assuming steady state: $N \cdot \underline{v} = \underline{0}$

$$\begin{pmatrix} 1 & 0 & -1 & 0 & 0 & 0 & 0 \\ 0 & 1 & 0 & -1 & -1 & 0 & 0 \\ 0 & 0 & 1 & 1 & 0 & -1 & 0 \\ 0 & 0 & 0 & 0 & 1 & 0 & -1 \end{pmatrix} \begin{pmatrix} 1 \\ 1 \\ v_3 \\ v_4 \\ v_5 \\ v_6 \\ 0.5 \end{pmatrix} = \begin{pmatrix} 0 \\ 0 \\ 0 \\ 0 \end{pmatrix}$$

Removing columns that corresponds to v_1, v_2, and v_7:

$$\begin{pmatrix} -1 & 0 & 0 & 0 \\ 0 & -1 & -1 & 0 \\ 1 & 1 & 0 & -1 \\ 0 & 0 & 1 & 0 \end{pmatrix} \begin{pmatrix} v_3 \\ v_4 \\ v_5 \\ v_6 \end{pmatrix} = - \begin{pmatrix} 1 & 0 & 0 \\ 0 & 1 & 0 \\ 0 & 0 & 0 \\ 0 & 0 & -1 \end{pmatrix} \begin{pmatrix} 1 \\ 1 \\ 0.5 \end{pmatrix}$$

$$= - \begin{pmatrix} 1 \\ 1 \\ 0 \\ -0.5 \end{pmatrix}$$

Let us bring the flux vector to the right and solve the system for v_3, v_4, v_5, and v_6:

$$\begin{pmatrix} v_3 \\ v_4 \\ v_5 \\ v_6 \end{pmatrix} = \begin{pmatrix} -1 & 0 & 0 & 0 \\ 0 & -1 & -1 & 0 \\ 1 & 1 & 0 & -1 \\ 0 & 0 & 1 & 0 \end{pmatrix}^{-1} \begin{pmatrix} -1 \\ -1 \\ 0 \\ 0.5 \end{pmatrix}$$

$$= \begin{pmatrix} -1 & 0 & 0 & 0 \\ 0 & -1 & 0 & -1 \\ 0 & 0 & 0 & 1 \\ -1 & -1 & -1 & -1 \end{pmatrix} \begin{pmatrix} -1 \\ -1 \\ 0 \\ 0.5 \end{pmatrix}$$

$$= \begin{pmatrix} 1 \\ +1 - 0.5 \\ 0.5 \\ +1 + 1 - 0.5 \end{pmatrix}$$

$$= \begin{pmatrix} 1 \\ 0.5 \\ 0.5 \\ 1.5 \end{pmatrix}$$

Drawing:

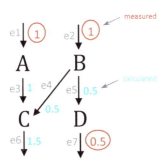

The kernel matrix K

Task a) Calculate the kernel matrix K of the following stoichiometric matrix N:

$$N = \begin{pmatrix} 1 & 1 & -2 \end{pmatrix}$$

The solution is:
With

$$N \cdot k_i = 0$$

we calculate:

$$\begin{pmatrix} 1 & 1 & -2 \end{pmatrix} \begin{pmatrix} k_{i1} \\ k_{i2} \\ k_{i3} \end{pmatrix} = 0$$

$$\Leftrightarrow 1k_{i1} + 1k_{i2} - 2k_{i3} = 0$$

In order to solve the system, we first need to determine the nullity:

$$r - \mathrm{rk}(N) = 3 - 1 = 2$$

We have an underdetermined system and 2 kernel vector are necessary to solve the system:
2 assumptions are necessary, $k_{i2} = c_2$ and $k_{i3} = c_3$:

$$k_{i1} + 1c_2 - 2c_3 = 0$$

$$\leftrightarrow k_{i1} = 2c_3 - c_2$$

$$\leftrightarrow k_{i1} = \begin{pmatrix} 2c_3 - c_2 \\ c_2 \\ c_3 \end{pmatrix}$$

Let us choose 2 non-identical values for c_2 and c_3: e.g. $c_2 = 1; c_3 = 0$

$$k_i = \begin{pmatrix} -1 \\ 1 \\ 0 \end{pmatrix}$$

e.g. $c_2 = 0; c_3 = 1$

$$k_i = \begin{pmatrix} 2 \\ 0 \\ 1 \end{pmatrix}$$

thus,

$$K = \begin{pmatrix} -1 & 2 \\ 1 & 0 \\ 0 & 1 \end{pmatrix}$$

Task b) Determine dead ends and unbranched reactions pathways, given the following Kernel matrix K:

$$K = \begin{pmatrix} -1 & 1 & 0 & 0 \\ 0 & 0 & -1 & 2 \\ -1 & 1 & 0 & 0 \\ 0 & 0 & 0 & 0 \\ 1 & 2 & -1 & 0 \\ 0 & 0 & -1 & 2 \\ -1 & 1 & 0 & 0 \\ -2 & 0 & 0 & 1 \end{pmatrix}$$

The solution is:
Dead ends: v_4
Unbranched reaction pathways:

- v_1, v_3, v_7

- v_2, v_6

◼ **Constraint-Based Modeling of an example network**
An example network is composed of substrates (S_1, S_2), intracellular metabolites (A, B, C) and the biomass (X).

- The following reactions v_1 to v_6 take place:

$$v_1 : S_1 \rightarrow A$$
$$v_2 : S_2 \rightarrow B$$
$$v_3 : C \rightarrow X$$
$$v_4 : A \rightarrow C$$
$$v_5 : B \rightarrow C$$
$$v_6 : A \rightleftharpoons B$$

- Reactions v_1, v_2, v_3, v_4, v_5 are irreversible

- For v_6, a maximal enzyme capacity is given: $v_6 \leqslant v_{6,max}$

- Due to the experimental setting, v_3 is fixed to $v_3 = D$

Questions:

Task a) Draw the biochemical network based on v_1 to v_6 and the irreversibility information.

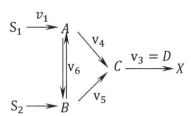

Task b) Formulate the balance equations for the intra-cellular metabolites A, B and C in steady state:

$$\dot{A} = v_1 - v_4 - v_6 = 0$$
$$\dot{B} = v_2 - v_5 + v_6 = 0$$
$$\dot{C} = v_4 + v_5 - v_3 = 0$$

Task c) Replace v_3 by D in the resulting equations. Keep v_1 and v_6 as variables and solve for v_2, v_4, and v_5:

$$A : v_4 = v_1 - v_6$$
$$C : v_4 + v_5 - D = 0$$
$$\Leftrightarrow v_1 - v_6 + v_5 - D = 0$$
$$\Leftrightarrow v_5 = D + v_6 - v_1$$
$$B : v_2 - D - v_6 + v_1 + v_6 = 0$$
$$\Leftrightarrow v_2 = D - v_1$$

Task d) Apply the irreversibility information of v_1 to v_5 and extract from there (several) inequality constraints

for the remaining variables v_1 and v_6:

$$v_1 \geq 0$$
$$v_2 = D - v_1 \geq 0$$
$$\rightarrow v_1 \leq D$$
$$v_3 = D$$
$$v_4 = v_1 - v_6 \geq 0$$
$$\rightarrow v_6 \leq v_1$$
$$v_5 = D + v_6 - v_1 \geq 0$$
$$\rightarrow v_6 \geq v_1 - D$$

Task e) Mark the possible solution space using a v_6 over v_1 plot using the obtained inequality constraints and the enzyme capacity $v_{6,max}$. Thereby distinguishing thereby two cases:

Task e.1) $v_{6,max} \geq D$

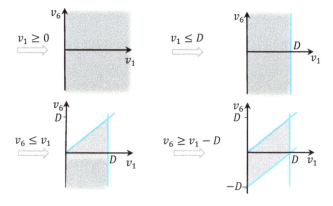

Help:
Case 1: $v_1 = D$
$v_6 \geq 0$ (fix the point)
Case 2: $v_1 = 0$
$v_6 \geq -D$ (fix the point)

Task e.2) $v_{6,max} < D$

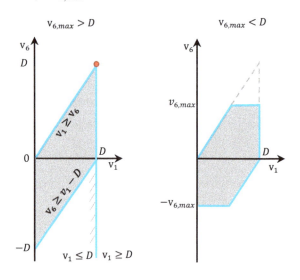

Task f)   Give the (unique) solution for the case $v_6 \geq D$ while optimizing for $v_6 \to max$.

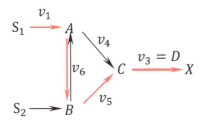

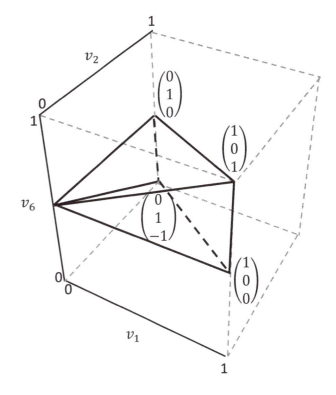

$$v_1 = v_3 = v_5 = v_6 = D$$
$$v_2 = v_4 = 0$$

Flux cone of an example network

From the previous system, we know that:

1. v_1 to $_5$ are irreversible

2. $| \; v_6 \leq v_{6,max}$

3. $v_3 = D$

$$
\begin{array}{lll}
v_1 & \geq 0 & v_1 & \geq 0 \\
v_2 & \geq 0 & v_2 & \geq 0 \\
v_3 & \geq 0 \Longrightarrow & v_1 + v_2 & \geq 0 \\
v_4 & \geq 0 & v_1 - v_5 & \geq 0 \\
v_5 & \geq 0 & v_2 + v_5 & \geq 0
\end{array}
$$

$$
\begin{pmatrix} \dot{c}_A \\ \dot{c}_B \\ \dot{c}_C \end{pmatrix} =
\begin{pmatrix}
1 & 0 & 0 & -1 & 0 & -1 \\
0 & 1 & 0 & 0 & -1 & 1 \\
0 & 0 & -1 & 1 & 1 & 0
\end{pmatrix}
\begin{pmatrix} v_1 \\ v_2 \\ v_3 \\ v_4 \\ v_5 \\ v_6 \end{pmatrix}
$$

with quasi-steady-state assumption for A, B, and C ($0 = Nv$):

$$
\begin{pmatrix} v_1 \\ v_2 \\ v_3 \\ v_4 \\ v_5 \\ v_6 \end{pmatrix} =
\begin{pmatrix}
1 & 0 & 0 \\
0 & 1 & 0 \\
1 & 1 & 0 \\
1 & 0 & -1 \\
0 & 1 & 1 \\
0 & 0 & 1
\end{pmatrix}
\begin{pmatrix} v_1 \\ v_2 \\ v_6 \end{pmatrix}
$$

Represented in 3D:

Notes

Chapter 3: The magic of change and how to find it

Thomas Sauter, Marco Albrecht

Motivation
The world around us is connected and changes all the time. But do we know where things are going? Are they on a path to endless infinity or do they find a condition where they are in balance with all their neighbors? In reality, we often see mysterious black boxes, and it is not clear why we get a particular outcome for a specific input. Sometimes the result depends on the history of movements. If this is the case, we see different final results for the same condition. In this block, we strive for answers in the realm of systems science. We will learn to model so-called systems in time, and we will ponder on how to bring them to desired states as quickly and precisely as possible. Likewise, we will learn how organisms develop an effective control of various regulatory biochemical processes forged by the forces of evolution. On this journey, we will also understand the beauty of connected things and what they teach us when we address biological problems, a perspective that make us better experimental biologists too.

Keywords
Modeling — Hysteresis — Feedback — Open and closed loop — Stability — Multi-stability — Phase plot

Contact: thomas.sauter@uni.lu. **Licence:** CC BY-NC

Contents

1. Lecture summary

In the previous two chapters, we studied systems in equilibrium and learned about powerful tools for the modeling of metabolic networks. However, the process

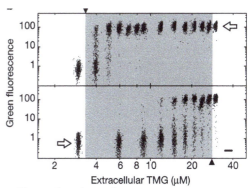

Figure 1. Example where the measurement seems mysterious. The GFP signal depends on the non-metabolizable lactose analogue thiomethyl-galactoside (TMG), but whether we increase TMG or decrease TMG remarkably changes the outcome, also referred to as hysteresis or multistability. What is behind this? See Figure 2. Source: [1]. Copyright © 2004, Macmillan Magazines Ltd.

of coming to a steady state has been assumed heretofore. In this chapter we will introduce some of the theoretical basics to treating and understanding biological systems which are dynamically changing over time and thus are no longer in steady state.

1.1 ■ Examples of non-linear dynamics
Let us start with a dynamic experiment on the populations of cells in Figure 1, which also shows that reality cannot always be described with basic linear tools. In this example, we see that the expression of genes relevant for the lactose metabolism (GFP signal) does not merely

https://doi.org/10.11647/OBP.0291.03

depend on the availability of the non-metabolizable lactate analogue (TMG). It also depends on the previous amount of TMG in the cellular environment. If 30 μg TMG were present at the beginning and this concentration was further decreased, gene expression for the lactose metabolism is measurable down to 5 μg TMG. But if, instead, the TMG concentration was increased starting with a low value of 2 μg TMG, gene expression will not set in until 15 μg TMG. Whether there will be lactose metabolism or not in the range from 5 to 15 μg TMG consequently depends on the history of the system, a phenomena known as hysteresis. Hysteresis can also be seen as a bistable state which can result in discrete switch-like outputs from continuous inputs.

Hysteresis increases the resistance to noise as it requires higher values of the input in order to switch to a particular state, as compared to the input needed to stay in a state. The switch also means that a transition is not continuously reversible after it has been triggered. The behavior is often hidden and encoded in unknown connections of biological elements. Here, the regulatory principles behind the observation in Figure 1 are well understood and illustrated in Figure 2.

Other examples of hysteresis can be studied in Figure 3 and 4, where the regulatory network structure (mutual molecule dependency) and the time behavior are depicted side by side. Take some time to understand how these schemes relate to each other. In Figure 4 we have an example from stem cell biology with the homeobox protein Nanog. Regulatory interconnections are not restricted to molecular biology. We might also have physiological and anatomical systems with regulatory interconnections. One example is the control of the blood sugar level by the liver and pancreas, illustrated in Figure 5. Another anatomical example can be found in Figure 6 on the adaptation of the eye. The restriction of the pupil compensates for disturbances with different light intensities. The control of this system can be drawn as a block diagram (see next section and Figure 7). Such regulatory circuits are also used in synthetic biology, within which scientist constructs or re-design existing biological systems for useful purposes. Let us explore the impact of some basic network motifs on the dynamics of a system next.

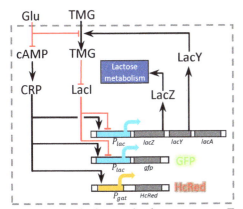

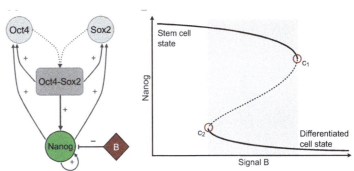

Figure 2. Underlying the hysteresis in Figure 1 is the specific regulatory network of the Lac operon model. LacY fascilitates the influx of TMG, which in turn inhibits the repressor. This again increases again the production of LacY, leading to a positive feedback loop and thus enabling bistability. Reproduced from: [1]. Copyright © 2004, Macmillan Magazines Ltd.

Figure 3. Stem cell differentiation network. Bifurcation diagram of transcription factor Nanog expression as a function of the levels of signal B. Solid lines represent stable states, dashed lines represent unstable states. The bistable region is shaded in gray. Bifurcations occur at signal intensities c1 and c2. Source: [2]. Copyright © 2015, The Company of Biologists Ltd.

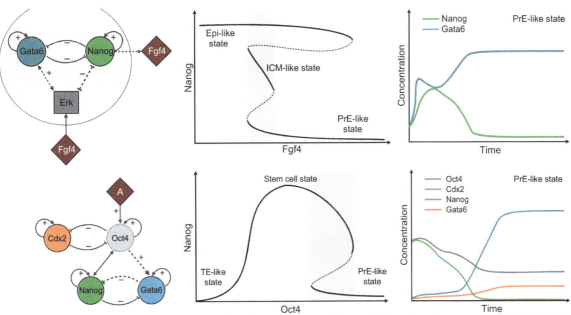

Figure 4. Stem cell differentiation network. A progenitor cell can differentiate into an epiblast cell (Epi), primitive endoderm (PrE; generating extra-embryonic membranes) and trophectoderm (TE; giving rise to extra-embryonic tissues), depending on the culture environment. This is a bifurcation diagram of transcription factor Nanog expression as a function of the levels of Fgf4 or Oct4 with different network motifs. Solid lines represent stable states, dashed lines represent unstable states. The bistable region is shaded in gray. TE-like state: stem cell switch with Nanog is off. PrE-like state: Gata6 expression is present but Nanog transcription is efficiently suppressed. ICM-like state: co-expression of Nanog and Gata6. Epi-like state: Nanog expression. Source: [2]. Copyright © 2015, The Company of Biologists Ltd.

Think about the following question: Could one use a linear regression approach between Nanog and Fgf4 or Oct4?

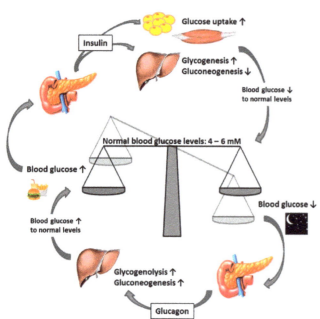

Figure 5. Blood sugar regulation. If the sugar concentration is too high in the blood, the pancreas releases insulin. Insulin makes tissue cells and the liver store or consume more sugar. If the sugar concentration is too low in the blood, the pancreas releases glucagon to cleave glycogen with subsequent up-regulation of blood sugar. Licence: CC BY-NC-ND 4.0 (Fair Use).

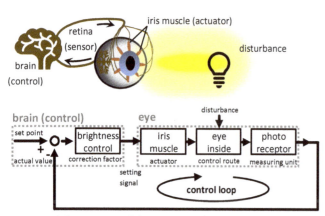

Figure 6. Adaptation of the eye to adjust to various light intensities. The light source disturbs the retina. The retina detects the light intensity and gives the actual value to the brain. The brain compares this value with the setpoint. If the light intensity is higher than acceptable, the brain gives the signal to reduce the light. If the light intensity is too low, the brain provides the signal to increase the light intensity. The actual adjustment is achieved via the pupil constriction by the iris muscle. Credit to Greyson Orlando, wikimedia. Fair Use.

1.2 ■ The impact of network motifs

We distinguish two main types of loops in networks: the feedback loop and the feed-forward loop (FFL).

Definition 1. Feedback: Processes which influence their own cause or input.

Feedback is a type of closed loop regulatory unit, as shown in Figure 7. Such control structures are widely used in engineering in many applications too—for example, to keep the output y at the desired level.

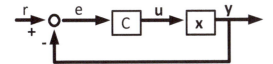

Figure 7. Block diagram of control loop (negative feedback).

A controller C uses the error e, which is the difference between the desired steady state r (set point) and the system output y, to adjust the control u. The control u acts on the system states x.

The feedback loop can encode negative and positive feedback, and occurs often in biology, as shown in Figure 10 with some examples. Negative feedback occurs frequently among repressors, which repress their own transcription. This allows a strong maximal promoter activity with a fast response, but at the same time strong repression once the desired product's steady state is achieved.

A variety of examples of network motifs have been identified in biological systems and analyzed. We can distinguish motifs with 1, 2, 3 and more than 3 nodes.

<u>One node:</u> In the *E. coli.* transcriptional network, 85% of the auto-regulative **self-edges** are negative and 10% are positive [3]. There are many cases where all edges start and come back to the same node.

<u>Two-node networks</u> with feedback occur in *e.g.* developmental networks and can be seen as memory elements. Once this motif is triggered, a mutual inhibition or activation is conserved and remains independent of further input under certain circumstances. A decision for the subsequent development has been made. For an example, see Figure 26 at the top with the toggle-switch motif. Coming to <u>three-node network motifs</u>, feed-forward loops are more than 30-fold enriched in the *E. coli* transcription network than random networks. Feed-forward loops can be split into **coherent FFL**, where the indirect path has the same overall sign as the direct path, and **incoherent FFL**, where the indirect path sign is the opposite to the sign of the direct path, shown in Figure 8.

Type I of both coherent and incoherent FFL are much more enriched in the *E. coli* network than the other types.

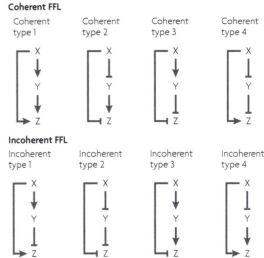

Figure 8. Feed-forward loops (FFL) split from one node and fall together at another node. Thereby information is transmitted in the forward direction via two parallel branches (pathways). The multiplication of the signs along the interactions of each branch gives the overall sign of a specific branch. A coherent FFL results in an end node which receives either only positive or only negative signals. An incoherent FFL end node receives both sign types. Source: [4]. Copyright © 1969, Nature Publishing Group.

The dynamic consequence of the **Type 1 feed-forward loop with AND gate** is shown in Figure 9 and Example 1. The output Z requires both activated X and activated Y. Once X is activated, Y begins to increase and is considered active as soon as it trespasses a certain threshold. If Y is active, it does not mean that Z is activated, as it requires maintained activation of X at the same time. This motif can be seen as a noise reducer in fluctuating environments. Only a signal that is present over a longer period can trigger a response, which arises then after a certain delay. An interruption of the input leads to instantaneous shutdown of the signal without delay. The **Type 1 coherent FFL with an OR gate** would activate Z instantaneously, and Y would have no relevance for the activation. Instead, the OR gate would make Y into a sustained activator of Z until it falls below the threshold. This motif would activate Z immediately and keep the response activation longer even if the original input was removed. The **Type 1 incoherent FFL with AND gate** would create a peak. We would see short activation of Z during long time-input activation. Once the level of Y trespasses the threshold, it inhibits the activation with different levels of strength, determined by the repression coefficient. The **Type 1 incoherent FFL with OR gate** would be activated instantaneously, but would accelerate the off-signal in Y. The acceleration increases with an increased repression coefficient.

Reminder: Logical operators such as OR and AND help us to define rules:

A	B	A AND B	A OR B
0	0	0	0
0	1	0	1
1	0	0	1
1	1	1	1

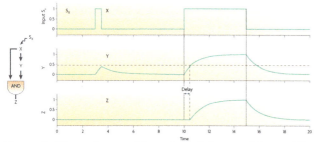

Figure 9. The coherent Type 1 feed-forward loop. The output Z shows delay after S_x addition in the input, but no delay after S_x removal. The network motif is a sign-sensitive filter, which responds only to persistent stimuli. Source: [4]. Copyright © 1969, Nature Publishing Group.

Example 1: Simulation of coherent Type 1 FFL

Referring to the dynamic feed-forward loop with AND gate and its dynamic behavior shown in Figure 9, we show here the underlying computational model which allows us to obtain such time course plots. This requires the formulation of balance equations for each of the modeled molecule concentrations (states), which include mathematical terms describing how the different effects change that specific concentration. This concept of balance equations will be explained in more detail and with more examples later in this chapter (Section 1.4, page 9) and in Chapter 4. So consider coming back to this example after reading more about balance equations.

This balance equation can then either be solved analytically—which we will discuss later in this chapter—or with the help of computer simulations. But these equations can also be solved with the help of computer simulations. Such simulations are usually used if the equation system is too complex to be solved analytically. Below, we give the computer model in the IQM toolbox format [5]. Other tools use similar but slightly different formats. This model is used within computational frameworks for numerical integration. This allows us to obtain usually very accurate approximations of the system behavior for a specific set of initial conditions and parameters. The parameters in this example

were arbitrarily chosen. The activation cutoff for Y matches the dashed line in Figure 9. Only above this threshold, Y transmits a positive signal (Yeff). The chosen initial conditions are matching the values of the states at $t = 0$.

```
********* MODEL NAME
FFL

********* MODEL STATES
d/dt(X) = k1*Sx - k1*X
d/dt(Y) = k2*X - k2*Y
d/dt(Z) = k3*Yeff*X - k3*Z

X(0) = 0
Y(0) = 0
Z(0) = 0

********* MODEL PARAMETERS
Sx = 0
k1 = 1
k2 = 1
k3 = 1
thrY = 0.5

********* MODEL VARIABLES
Yeff = piecewiseIQM(Y-thrY, Y>thrY, 0)
```

All motifs are possible considering networks with more than 3 nodes. It is imaginable that one element in a signaling branch might inhibit the first element of that branch in a negative feedback loop. This section builds upon the work of Uri Alon. His book is a classic in systems biology and highly recommended to help you understand regulatory network motifs [3].
We will revisit feedbacks again within this Chapter (Example 9 & Figures 23/24/25 and the respective text), once we have learned about characteristics of systems and useful mathematical approaches which will help us to formally analyze such motif structures.

For further deepening of knowledge around this topic we also recommend:
YouTube: Network motifs

Summary: Mode of action of network motifs

positive feedback	= self-potentiating
	→ (often) destabilizing
	→ bifurcations
	→ bistabilities
	→ switching behavior
	→ hysteresis
negative feedback	= self-degrading
	→ stabilizing
	→ oscillating
	→ regulating

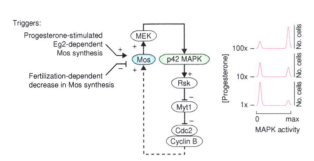

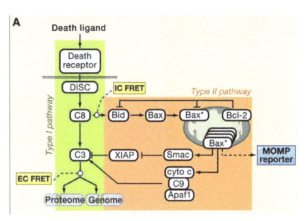

(a) **Naturally occurring bistable system:** The p42 MAPK cascade of a frog oocyte depends on the progesterone stimulation. The more progesterone, the more the cells shift to MAPK active differentiation. Source: [6]. Copyright © 2002, Elsevier Science Ltd.

(b) **Switch-like activation and bistability** in the apoptosis signaling network. Source: [7]. Copyright © 2011, Elsevier Inc.

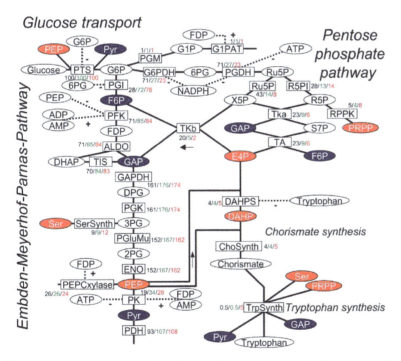

(c) **Negative tryptophan feedback:** Numbers indicate molar fluxes normalized to the glucose uptake flux (100). Each of the 3 numbers belong to another model variant. Source: [8]. Copyright © 2004, Elsevier Inc.

Figure 10. Naturally occurring feedback in biology: (a) Positive feedback on the genetic level. (b) Positive feedback on the protein level. (c) Negative feedback on the metabolic level.

1.3 ■ What is a system?

Next, we are moving into the formal description of such networks as systems.

Definition 2. System: A system is a group of interdependent items which are interacting with each other in a way that forms an integrated whole.

One can also say: a system is a set of interrelated objects (elements, parts) which all have certain general properties [9].

1. The system fulfils a certain function, *i.e.* it can be defined by a **system purpose** recognizable by an observer.

2. It has a characteristic constellation of (essential) **system elements** and an (essential) **system structure** which determines its function, purpose, and identity.

3. It loses its **identity** if its integrity is destroyed. A system is therefore **not divisible**, *i.e.* the system purpose can no longer be fulfilled if one or several (essential) elements are removed.

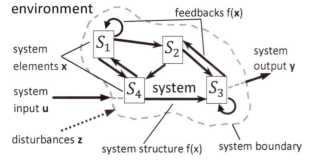

Figure 11. A system consists of system elements and the connections between them [9].

As a side note, the system structure might be (partially) hidden, *i.e.* we do not know what is underlying the observed system dynamic. We might refer to the system, then, as a **black box**—and we could try to decipher the system behavior by analysis of the relationship of inputs *u* and outputs *y*, as illustrated in Figure 12.

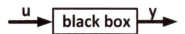

Figure 12. The black box concept. We know little about the system, but we can learn about its behavior from the input *u* to the output *y*.

Let us look at some examples. A sand pile is not a system, as after the removal of a sub-part it is still a

sand pile. One grain of sand is also not a system, as it is only one element. A chair or a building is a system because the removal of one particular element destroys its integrity. These are static systems. Dynamic systems show changes over time. We do not have to restrict ourselves to observable behaviors. Relevant state variables are of functional importance but not necessarily observable. Furthermore, the non-linearity of biological systems (non-additivity) is prevalent.

Linear: $v[S_1] + v[S_2] = v[S_1 + S_2]$

Non-linear: $v[S_1] + v[S_2] \neq v[S_1 + S_2]$

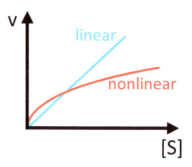

As we have seen so far already, non-linear dynamics can lead to a variety of phenomena like:

- oscillations

- amplification

- adaptation (see Figure 13)

- switching, bistability, hysteresis (see Figures 1,3,4).

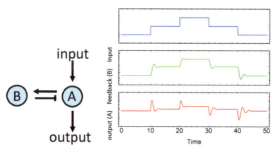

Figure 13. Near-perfect adaptation to varying input levels obtained from negative feedback. Other examples and source: [10]. Copyright © 2016, Elsevier Inc.

See also
YouTube: The beauty of a chaotic double pendulum

We will explore these important behaviors on the next pages, starting with the introduction of a mathematical approach for describing dynamics over time.

1.4 ■ How to describe change mathematically

The change of a variable is described by a derivative—a concept from calculus. The most widely used mathematical approach to capture such changes in models is done with **Ordinary Differential Equations** (ODE). Such ODEs have a dependent variable y which is a function $y(t)$ of $\underline{1}$ independent variable such as time t. The derivatives of the dependent variable are found with respect to the independent variable $\frac{dy}{dt}$. If the independent variable is time, we can use the dot above the dependent variable $\frac{dy}{dt} = \dot{y}$ to indicate derivatives, otherwise we use a prime symbol: y'. The dependent and independent variable appear in the **explicit**:

$$\dot{y} = f(t, y)$$

or **implicit** equation form:

$$f(t, y, \dot{y}) = 0.$$

Important: In the context of models of biological systems, often only 1st order ODEs are applied, as illustrated in Example 2. We have therefore included an introduction to these cases only. ODEs might also contain higher derivatives, *i.e.* derivatives of derivatives. Furthermore, **Partial Differential Equations** (PDE) have a dependent variable y, which is a function $y(z_1, z_2, \ldots, z_n)$ of at least 2 independent variables z_i, such as time (*e.g.*: $z_1 = t$) and 3 room coordinates (*e.g.*: $z_2, z_3, \& z_4$). While an ODE is frequently used for processes over time, a PDE is often used to describe processes over time and space. These advanced cases are not discussed here but are the subject of further reading if you are interested.

A simple ODE model can be studied in Example 2 together with Figure 14.

Example 2: Simple ODE for biomass growth

Another example of what x and its time derivative $\dot{x}$ might represent is biomass and its growth.

$x \ldots$ population size

$\dot{x} \ldots$ growth rate

$\mu = \frac{\dot{x}}{x} \ldots$ specific growth rate

With the assumption $\mu = const$, we can write the following ODE:

$$\dot{x} = \mu x$$

with initial condition $x(t = 0) = x_0$ and

$\mu > 0$: exponential growth

$\mu < 0$: asymptotic decay

---

visualized in Figure 14. The calculation of the time response $x(t)$ is:

$$\frac{dx}{dt} = \mu x \qquad |\cdot dt : x$$

$$\frac{1}{x} dx = \mu dt \qquad |\int$$

$$\int_{x_0}^{x} \frac{1}{x} dx = \mu \int_{0}^{t} 1 dt$$

$$[ln(x)]_{x_0}^{x} = \mu(t - 0)$$

$$ln(x) - ln(x_0) = \mu t$$

$$ln\left(\frac{x}{x_0}\right) = \mu t \qquad |e \text{ to both sides}$$

$$\frac{x}{x_0} = e^{\mu t}$$

$$x(t) = x_0 e^{\mu t}$$

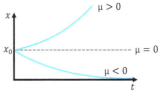

Figure 14. Population growth x depending on time and the specific growth rate μ. See Example 2.

A set of non-linear ODEs of n-th order can be written as:

$$\frac{dx_i}{dt} = \dot{x}_i = f_i(x_1, \ldots, x_n, p_1, \ldots, p_n, t) \qquad , i = 1 \ldots n$$

and in vector notation as:

$$\frac{d\mathbf{x}}{dt} = \dot{\mathbf{x}} = \mathbf{f}(\mathbf{x}, \mathbf{p}, t)$$

with the independent variable t, the time-dependent states $\mathbf{x} = (x_1, \ldots, x_n)^T$, mostly time-invariant parameters $\mathbf{p} = (p_1, \ldots, p_n)^T$, and the functions $\mathbf{f} = (f_1, \ldots, f_n)^T$. Although the system states are time-dependent, they should still be linearly independent of each other. The **order of an ODE** equals the number states within the set of 1st order ODEs.

Reactions depend on changing substrate concentrations
In the last chapter, we considered constant substrate concentration in the steady state:

$$\dot{\mathbf{S}} = \mathbf{N} \cdot \mathbf{v} = \mathbf{0}$$

and we looked mainly at the fluxes $\mathbf{v}$, which can also be dissected into forward v_f and backward v_b reaction fluxes. If the substrate concentrations are constant, it is possible to combine these with the constant parameters to a flux value for a given reaction.

With the study of the time behavior of each substrate's concentrations in our system with:

$$\dot{S} = N \cdot v$$

we cannot do this anymore. We have to dissect the flux vectors into constant parameters k and now variable substrate concentrations S_i. Therefore, we usually use the very important **Law of Mass Action** formulated by Waage[1] and Guldberg[2] in 1864 [11]. For each reaction j, we can define the concentrations of substrates S_i and products P_i:

$$\sum_{\alpha} m_{j\alpha} S_{j\alpha} \underset{k_{-j}}{\overset{k_j}{\rightleftharpoons}} \sum_{\beta} m_{j\beta} P_{j\beta}$$

with indexes α and β being subsets of the metabolite indexes $i = 1, \ldots, m$ and molecularities $m_{j\alpha}, m_{j\beta}$ for reaction j and participating substances $\alpha, \beta \in i$. The molecularity is often 1 or 2.

We can say that the **reaction rate** is proportional to the **probability of collision** of the reactants. The probability of collision, in turn, is proportional to the concentration of reactants to the power of molecularity (stoichiometric coefficient). The general mass action kinetics for chemical reaction networks is:

$$v_j = v_{j,f} - v_{j,b} = k_{+j} \prod_{\alpha} S_{j\alpha}^{m_{j\alpha}} - k_{-j} \prod_{\beta} P_{j\beta}^{m_{j\beta}}$$

with the product of sequence operator $\prod$ which indicates that the following elements are multiplied (same principle as for the operator $\sum$ to sum up the following terms). The equilibrium constant:

$$\Rightarrow K_{j,eq} = \frac{k_{+j}}{k_{-j}} = \frac{\prod_{\beta} P_{j\beta,eq}^{m_{j\beta}}}{\prod_{\alpha} S_{j\alpha,eq}^{m_{j\alpha}}}$$

for reaction j can be obtained if the substances are in equilibrium $v_{j,f} = v_{j,b}$. In contrast to Chapter 2, here we do not have a dynamic equilibrium of metabolites' production and consumption.

For the biochemical reaction:

$$S_1 + S_2 \underset{k_{-1}}{\overset{k_1}{\rightleftharpoons}} 2P$$

we get the **net reaction rate:**

$$v = v_f - v_b = k_1 \cdot S_1 \cdot S_2 - k_{-1} \cdot P^2$$

with S_1, S_2 and P denoting the respective concentrations and k_1 and k_{-1} being the rate constants of the forward and backward reactions.

One can also derive the equations directly from the reaction schemes without using the stoichiometric matrix. This might be error-prone for larger biochemical systems and molecularities of 2 and above. For the given scheme, the ODE system is:

$$\dot{S}_1 = -k_1 S_1 S_2 + k_{-1} P^2$$
$$\dot{S}_2 = -k_1 S_1 S_2 + k_{-1} P^2$$
$$\dot{P} = 2k_1 S_1 S_2 - 2k_{-1} P^2$$
$$= \underbrace{2}_{N} \cdot \underbrace{(k_1 S_1 S_2 - k_{-1} P^2)}_{v}$$

This modeling principle can equivalently by applied for protein-binding reactions where the parameters k represent the affinity between proteins—for example, between ligand and receptor:

$$L + R \underset{k_{-1}}{\overset{k_1}{\rightleftharpoons}} LR$$

with the net reaction rate (from left to right):

$$v = v_{bind} - v_{unbind} = k_1 \cdot L \cdot R - k_{-1} \cdot LR$$

We could, for example, use this equation to see how much ligand-receptor complex LR will be present for certain initial concentrations of ligand L and receptor R. High affinity with a large constant $K_a = k_1/k_{-1}$ shifts the equilibrium towards the complex, and low affinity shifts this equilibrium to the monomers. After we have modeled a reaction network, an equilibrium can be found for the whole system. Trying to derive such an equilibrium for many coupled reactions by intuition alone might be too challenging. Thus, we need mathematical models. The ODE modeling framework gives the highest quality of mechanistic insight among all tools in systems biology—but only if we have enough data to adjust this kind of model properly.

---

[1] Norwegian chemist Peter Waage (1833–1900).
[2] Norwegian mathematician and chemist Cato Maximilian Guldberg (1836—1902).

1.5 ■ What is a steady state?

A steady state is characterized by the absence of change in the variable values while time progresses.

Definition 3. Steady state: In systems theory, a system or a process is in a steady state if the variables (called state variables) which define the behavior of the system or the process are not changing over time.

This definition is referring to macro-scale changes over time. Back reactions equal forward reactions at the molecular scale in a closed system. In an open system, a dynamic equilibrium occurs in which fluxes and reactions are constant. We have a steady state if all derivatives of all state variables x_i become zero:

$$\dot{x}_i = \ddot{x}_i = \dddot{x}_i = \ldots = 0$$

for all i. Moreover, specifically for the 1st-order systems we consider in this chapter, we have a steady state if the (1st-order) derivatives of all state variables x_i become zero:

$$\dot{x}_i = 0$$

for all i.

Consider the following Examples 3, 4, and 5:

Example 3: Balance equation in steady state

$$\frac{d\Phi}{dt} = \dot{\Phi} = J - P$$

Steady state?

$$\dot{\Phi} = J - P = 0$$
$$\Rightarrow J = P$$

Example 4: One steady state of a linear system

The derivatives of all states shall be zero:

$$\dot{x} = y + 2x - 2 = 0$$
$$\dot{y} = y - x + 2 = 0$$

Subsequently, we mark all state values with a star * to indicate that these are steady-state values and re-order the equations above as follows:

$$y^* = 2 - 2x^*$$
$$y^* = x^* - 2$$

Eliminating y^* by setting both of the right-hand sides to be equal delivers:

$$2 - 2x^* = x^* - 2$$

which gives the steady-state solution:

$$x^* = \frac{4}{3}, \quad y^* = -\frac{2}{3}$$

Example 5: Steady states of a non-linear system

We work on a non-linear system and will thereby often obtain multiple steady states, like in this example:

$$\dot{x} = (y + 2x - 2)x = 0$$
$$\dot{y} = (y - x + 2)y = 0$$

This set of equations is already in the factored form. 1 of the 2 factors in each single equation must be zero so that the whole term is zero. We have 2 equations with 2 factors each. Thus, we expect 4 steady states, whereby a subset can be the same. To simplify our approach, we will look at different cases.

Case 1

Let's say the first variable x^* shall be zero:

$$\dot{x} = (y + 2x - 2)\overbrace{x}^{x^*_{1,2}=0} = 0$$
$$\dot{y} = \underbrace{(y - x + 2)}_{y^*_2}\underbrace{y}_{y^*_1} = 0$$

Then Equation 1 is zero as well. The second equation is then:

$$0 = (y^* + 2)y^*$$

which results in two allocated y^* values:

$$y^*_1 = 0, \quad y^*_2 = -2$$

with which we get our first 2 steady states $(0,0)$ and $(0,-2)$.

Case 2

We now take the other factor $(y + 2x - 2)$ in the first equation and set it to zero:

$$\dot{x} = \overbrace{(y + 2x - 2)}^{\overset{!}{=}0 \to y^*_{3,4}} x = 0$$
$$\dot{y} = \underbrace{(y - x + 2)y}_{x^*_{1,2}} = 0$$

Therefore, y^* has to be:

$$y^* = 2(1 - x^*)$$

with which the second equation becomes:

$$0 = 2(1-x^*)(4-3x^*)$$

which results in:

$$x_3^* = 1, \quad x_4^* = \frac{4}{3}$$

and two allocated y^* values with which we get our last 2 steady states $(1,0)$ and $(\frac{4}{3}, -\frac{2}{3})$.

Case 3 (redundant)
Ansatz: $y^* = 0$
with the solutions $(1,0)$ and $(0,0)$.

Case 4 (redundant)
Ansatz: $y^* = x^* - 2$
with the solutions $(0,-2)$ and $(\frac{4}{3}, -\frac{2}{3})$.

1.6 ■ Stability theory: Idle state or explosion

Stability is a system-intrinsic property and is characterized without external input $u = 0$. A non-linear system can have several steady states. The system's steady state might be slightly deflected by δx:

$$x_0 = x^* + \delta x$$

and the subsequent behavior indicates whether a system is stable, unstable, or metastable—as illustrated in Figure 15.

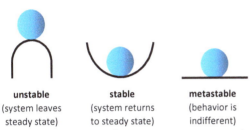

unstable
(system leaves steady state)

stable
(system returns to steady state)

metastable
(behavior is indifferent)

Figure 15. To get an initial intuition, imagine a ball on different surfaces. If you poke it a bit, what will happen?

To test this, we linearize around the steady state of the example system $\dot{x} = mx$ and can see by the slope m whether the steady state is stable or not. Linearization is scrutinized in the mathematics section. Ponder on Figure 16. The linearization:

$$\Delta \dot{x} = J \Delta x$$

with the Jacobian matrix J is explained in the mathematics section. The formal approach employs the exponential function starting from the system equation:

$$\dot{x} = mx$$

with the ansatz $x(t) = e^{st}$ and its derivative $\dot{x}(t) = s \cdot e^{st}$, and consequently we transform the system into the **Laplace Domain**. It results in:

$$se^{st} = me^{st} \Rightarrow s = m$$

with s being the Eigenvalue of system $\dot{x} = mx$. See also Figure 17.

The characteristic polynomial and Eigenvalues for higher order systems

In general, the characteristic polynomial allows us to determine the Eigenvalues of the system and thereof the stability of the steady states.

An ordinary differential equation in the non-matrix representation

$$\overset{(4)}{x} + \ldots + a_2\ddot{x} + a_1\dot{x} + a_0x = b_0u$$

is the base for the characteristic polynomial without consideration of the control ($u \overset{!}{=} 0$). We use the Laplace

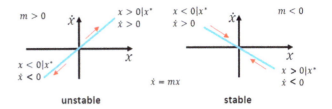

Figure 16. Assessing stability with an $\dot{x}$ over x plot. For the unstable steady state, if x is deviating in the positive direction from steady state, the resulting derivative will also be positive and thus make x even more positive—and *vice versa* for negative deviations. Contrarily, for the stable steady state the derivative (change) will be negative if x is positive and positive if x is negative. This will outbalance deviations and stabilize the steady state.

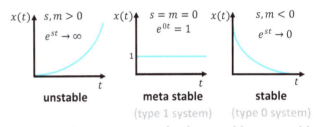

Figure 17. Behavior over time for the unstable, metastable, and stable system of the form $\dot{x} = mx$ with different parameter values m.

ansatz:

$$
\begin{aligned}
x &= e^{st} \\
\dot{x} &= se^{st} \\
\ddot{x} &= s^2 e^{st} \\
\dddot{x} &= s^3 e^{st} \\
\overset{(4)}{x} &= s^4 e^{st}
\end{aligned}
$$

to transform our problem into the frequency domain and get:

$$s^n + \ldots + a_2 s^2 + a_1 s + a_0 = 0$$

Example 6: 2nd-order system

$$\ddot{x} + a_1 \dot{x} + a_0 x = 0$$

ansatz:

$$x = e^{st}$$
$$\dot{x} = s e^{st}$$
$$\ddot{x} = s^2 e^{st}$$

$$s^2 e^{st} + a_1 s e^{st} + a_0 e^{st} = 0 \qquad | : e^{st} (\neq 0)$$
$$s^2 + \underbrace{a_1}_{tr} s + \underbrace{a_0}_{det} = 0$$

$$\Rightarrow s_{1,2} = \frac{-a_1 \pm \sqrt{a_1^2 - 4a_0}}{2}$$

The linear $\dot{x} = Ax$ or linearized system $\dot{x} = Jx$ in the matrix form can also be treated with the same ansatz:

$$x = I e^{st}$$
$$\dot{x} = s I e^{st}$$

and the identity matrix I to get:

$$s I e^{st} = A I e^{st} = A e^{st}$$
$$(sI - A) e^{st} = 0$$
$$(sI - A) x = 0$$

We only obtain non-trivial solutions ($x \neq 0$) of the homogeneous system if we look for:

$$\det(sI - A) = 0,$$

which is another form of the **characteristic polynomial**. However, if the system matrix is a square matrix, and the rank equals the number of variables ($\mathrm{rk}(A) = n$), we only obtain the trivial solutions ($x = 0$).

Eigenvalues of the 2x2 matrix (fast equation)
The characteristic polynomial of a 2nd-order system ($n = 2$) is introduced in Example 7 and can be obtained by the following procedure:

$$\dot{x}_1 = a_{11} x_1 + a_{12} x_2$$
$$\dot{x}_2 = a_{21} x_1 + a_{22} x_2$$
$$\Rightarrow$$

$$\underbrace{\begin{pmatrix} \dot{x}_1 \\ \dot{x}_2 \end{pmatrix}}_{\dot{x}} = \underbrace{\begin{pmatrix} a_{11} & a_{12} \\ a_{21} & a_{22} \end{pmatrix}}_{A} \underbrace{\begin{pmatrix} x_1 \\ x_2 \end{pmatrix}}_{x} = 0$$

with:

$$x = I e^{st}$$
$$\dot{x} = s I e^{st}$$

we get:

$$s x_1 = a_{11} x_1 + a_{12} x_2$$
$$s x_2 = a_{21} x_1 + a_{22} x_2$$
$$\Rightarrow$$

$$s \underbrace{\begin{pmatrix} x_1 \\ x_2 \end{pmatrix}}_{\dot{x}} = \underbrace{\begin{pmatrix} a_{11} & a_{12} \\ a_{21} & a_{22} \end{pmatrix}}_{A} \underbrace{\begin{pmatrix} x_1 \\ x_2 \end{pmatrix}}_{x} = 0$$

$$\underbrace{\left(s \begin{pmatrix} 1 & 0 \\ 0 & 1 \end{pmatrix} - \begin{pmatrix} a_{11} & a_{12} \\ a_{21} & a_{22} \end{pmatrix} \right)}_{=0} \underbrace{\begin{pmatrix} x_1 \\ x_2 \end{pmatrix}}_{\neq 0} = 0$$

$$|sI - A| = \left| s \begin{bmatrix} 1 & 0 \\ 0 & 1 \end{bmatrix} - \begin{bmatrix} a_{11} & a_{12} \\ a_{21} & a_{22} \end{bmatrix} \right|$$

$$= \left| \begin{bmatrix} s & 0 \\ 0 & s \end{bmatrix} - \begin{bmatrix} a_{11} & a_{12} \\ a_{21} & a_{22} \end{bmatrix} \right|$$

$$= \left| \begin{matrix} s - a_{11} & -a_{12} \\ -a_{21} & s - a_{22} \end{matrix} \right| = 0$$

$$(s - a_{11})(s - a_{22}) - a_{12} a_{21} = 0$$
$$s^2 - a_{11} s - a_{22} s + a_{11} a_{22} - a_{12} a_{21} = 0$$
$$s^2 - \underbrace{(a_{11} + a_{22})}_{tr(A)} s + \underbrace{a_{11} a_{22} - a_{12} a_{21}}_{\det A} = 0$$
$$s^2 - tr(A) s + \det A = 0$$

$$s_{1,2} = \frac{tr(A) \pm \sqrt{tr^2(A) - 4 \det(A)}}{2}.$$

s_1, s_2 can be real numbers but also complex numbers of the form $s = Re(s) + i Im(s)$ with $i = \sqrt{-1}$. If the number is complex, the system has oscillating behavior. See Figure 18 for different Eigenvalue pairings and the consequential systems behavior. Additionally, we can also classify the systems according to the Final Value Theorem (FVT). Eigenvalues on the imaginary axis and the right half of the plane indicate that the system does not converge to a value and is thus not classified with the FVT. We classify the system type according to the number of Eigenvalues in the origin. If the real part of the Eigenvalues is on the left-hand side, we have a Type 0 system. If one Eigenvalue is on the origin and the rest is in the left-half plane, we have a Type 1 system that is going to have a real finite number, as shown in Figure 17 middle. With each additional Eigenvalue at the origin, we increase the system type order and get unstable behavior.
YouTube: Final Value Theorem (up to 8:17).

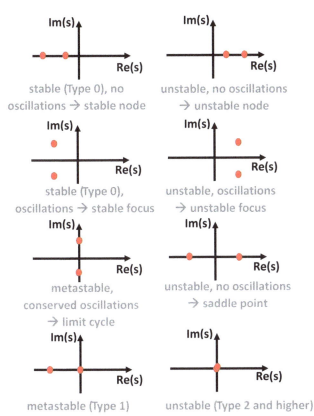

Figure 18. Eigenvalues in the frequency domain are shown in the s-plane. Each Eigenvalue has a real and imaginary part: $s = \sigma + i\omega$ with $Im(s) = i\omega$, and $Re(s) = \sigma$

Example 7: Eigenvalues and stability

a)

$$\begin{pmatrix} \dot{x}_1 \\ \dot{x}_2 \end{pmatrix} = \begin{pmatrix} 3 & -2 \\ 1 & 5 \end{pmatrix}\begin{pmatrix} x_1 \\ x_2 \end{pmatrix} + \begin{pmatrix} 1 \\ 0 \end{pmatrix} u$$

We search for the Eigenvalues with:

$$\det(A - sI) = 0$$

$$\det\begin{pmatrix} 3-s & -2 \\ 1 & 5-s \end{pmatrix} = 0$$

$$(3-s)(5-s) + 2 = 0$$

$$s^2 - 8s + 17 = 0$$

$$s_{1,2} = -\frac{-8}{2} \pm \sqrt{\left(\frac{-8}{2}\right)^2 - 17}$$

$$\Rightarrow s_{1,2} = 4 \pm i \quad \rightarrow \text{unstable focus}$$

b)

$$\begin{pmatrix} \dot{x}_1 \\ \dot{x}_2 \end{pmatrix} = \begin{pmatrix} -1 & 4 \\ 0 & -3 \end{pmatrix}\begin{pmatrix} x_1 \\ x_2 \end{pmatrix}$$

$$\Rightarrow s_1 = -1; s_2 = -3 \quad \rightarrow \text{stable node}$$

c)

$$\begin{pmatrix} \dot{x}_1 \\ \dot{x}_2 \end{pmatrix} = \begin{pmatrix} 3 & 8 \\ 1 & 1 \end{pmatrix}\begin{pmatrix} x_1 \\ x_2 \end{pmatrix}$$

$$\Rightarrow s_1 = 5; s_2 = -1 \quad \rightarrow \text{unstable saddle}$$

d)

$$\begin{pmatrix} \dot{x}_1 \\ \dot{x}_2 \end{pmatrix} = \begin{pmatrix} -1 & 3 \\ -1 & 1 \end{pmatrix}\begin{pmatrix} x_1 \\ x_2 \end{pmatrix}$$

$$\Rightarrow s_{1,2} = 0 \pm \sqrt{2}i \quad \rightarrow \text{metastable limit cycle}$$

Stability classification

We can classify the stability properties via the Eigenvalues, as summarized in Figure 18. For 2x2 matrices, we can also use the trace and the determinant of the system matrix as shown in the Poincaré diagram in Figure 21:

$$\lambda = \frac{\text{tr}(A) \pm \sqrt{\text{tr}^2(A) - 4\det(A)}}{2}.$$

with the distance between Eigenvalues:

$$\Delta =: \sqrt{\text{tr}^2(A) - 4\det(A)} = 0$$

$$\det(A) = \frac{1}{4}\text{tr}^2(A)$$

to get the threshold above which we gain imaginary numbers.

- $\lambda_1, \lambda_2 < 0$ then $\det(A) > 0, \text{tr}(A) < 0 \rightarrow$ stable node
- $\lambda_1, \lambda_2 > 0$ then $\det(A) > 0, \text{tr}(A) > 0 \rightarrow$ unstable node
- $\lambda_1 < 0, \lambda_2 > 0$ then $\det(A) < 0 \rightarrow$ saddle node with one stable and one unstable direction
- $Re(\lambda_1) = Re(\lambda_2) < 0$ with no vanishing imaginary part. Then $\det(A) > 0, \text{tr}(A) < 0, \text{tr}^2(A) < 4\det(A) \rightarrow$ stable spiral
- $Re(\lambda_1) = Re(\lambda_2) > 0$ with no vanishing imaginary part. then $\det(A) > 0, \text{tr}(A) > 0, \text{tr}^2(A) < 4\det(A) \rightarrow$ unstable spiral

Slopefield

A slopefield describes the slope at each point in the phase plot and thereby is an informative visual representation of the system dynamics.

YouTube: Slope fields

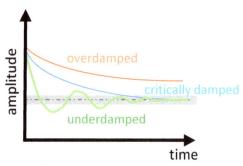

Figure 19. An ODE system can show various behavior depending on the damping ratio. The higher the damping ratio is, the more the oscillatory behaviour is reduced. See Example 8.

Example 8: Damping

For a deeper understanding of the origin of real and imaginary Eigenvalues and the respective system behavior, it is beneficial to look at damping, as shown in Figure 19. Damping reduces the impact of the oscillatory behavior. In physical systems it is achieved by the dissipation of energy stored in the oscillatory system. Let's assume a system has the following characteristic equation:

$$s^2 + 2s + \zeta = 0.$$

The quadratic equation can be solved with the p-q-equation:

$$s_{1,2} = -1 \pm \sqrt{1 - \zeta}$$

We see that the first term indicates negative Eigenvalues at the first glimpse. This remains the case as long $0 \leq \zeta \leq 1$. If we have $\zeta < 0$ we have one positive Eigenvalue and thus an unstable saddle. If $\zeta > 1$, we obtain a negative value below the square root and the term turns into the imaginary part $\pm \alpha i$ of the complex number s. Consequently, we have oscillatory behaviour. Thus ζ has a damping function with:

$0 \leq \zeta \leq 1$ overdamped

$\zeta = 1$ critically damped

$\zeta > 1$ underdamped

The smooth transition within the p-q equation and the symbol $\pm$ explains why the imaginary parts of the complex Eigenvalues arise pairwise with the same real part in Figure 18. This only applies to simple systems with a 2x2 system matrix A.

1.7 ■ Phase portrait: How elements relate to each other

Let's assume we have a stable system with two elements x_1 and x_2.

$$\dot{x}_1(t) = f_1(x_1(t), x_2(t))$$
$$\dot{x}_2(t) = f_2(x_1(t), x_2(t))$$

We do not only want to know how the elements change over time. We also want to know how the elements relate to each other:

$$\frac{dx_2(t)}{dx_1(t)} = \frac{f_2(x_1(t), x_2(t))}{f_1(x_1(t), x_2(t))}$$

by making use of the relationship:

$$\frac{\dot{x}_2(t)}{\dot{x}_1(t)} = \frac{\frac{dx_2}{dt}}{\frac{dx_1}{dt}} = \frac{dx_2}{dt}\frac{dt}{dx_1} = \frac{dx_2}{dx_1}.$$

Put in simple terms, the time derivative has been cancelled out and we have no time axis anymore, as illustrated in Figure 20. But time is still present, because at each time point and for each state magnitude x_2 has an allocated state magnitude x_1. For the whole time, it could be that the system has passed different magnitudes of x_2 at a particular magnitude of x_1. Moreover, there might be several different values for x_1 at the magnitude x_2. But let us go back to the start. You can now choose which initial values you would like to start with $[x_1(t=0), x_2(t=0)]$. From this point on, the system will move in the phase plot until it reaches a stable state where it will remain forever $[x_1(t \to \infty), x_2(t \to \infty)]$. The path which the system takes until it arrives a stable point is called a trajectory.

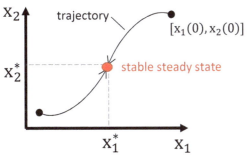

Figure 20. Phase portrait. The axes represent the magnitude of the elements x_1, x_2. The trajectories start from $[x_1(t=0), x_2(t=0)]$ and move over time to a stable steady state $[x_1(t \to \infty), x_2(t \to \infty)]$. In this way, we can plot our system as the change of element x_2 over the change of the element x_1. One could also decide to show the relationship the other way around, *e.g.*, the phase plot as the change of x_1 over the change of x_2.

Definition 4. Trajectory: Set of all state points (x_1, x_2) in the state plane starting from t_0 or $t = 0$.

Interestingly, not all initial points might lead to the same steady state. Depending on your initial point, the trajectories might be shaped differently and might land at different steady states, given that you have multiple stable points. Thus, it is interesting to look at several trajectories which start at different initial points. If we want to draw a phase plot by hand, we would greatly benefit from orientation lines. Therefore, we determine asymptotes with the most simple cases as being:

- Vertical asymptotes (vertical tangents)

- Horizontal asymptotes (horizontal tangents)

- Oblique asymptotes.

What horizontal and vertical means depends on which element you want to study in dependence on the other. We remain in the general equations shown above (see also Figure 20). We obtain the horizontal and vertical tangents with the following *ansatz*:

- Horizontal tangent ("no change in the x_2-direction"):

$$f_2(x_1, x_2) = 0 \tag{1.1}$$

- Vertical tangent ("no change in the x_1-direction"):

$$f_1(x_1, x_2) = 0 \tag{1.2}$$

For the oblique asymptote, we introduce the general linear equation into our phase plot equation:

$$\frac{d(x_2 = mx_1)}{dx_1} = m = \frac{f_2(x_1, (x_2 = mx_1))}{f_1(x_1, (x_2 = mx_1))}.$$

The obtained values for m are set back in the linear equation $x_2 = mx_1$ and represent the oblique asymptotes. For a deeper discussion of this topic, please refer to the exercises at the end of this chapter.

Definition 5. Nullclines: Curves on which the trajectories have the derivative 0.

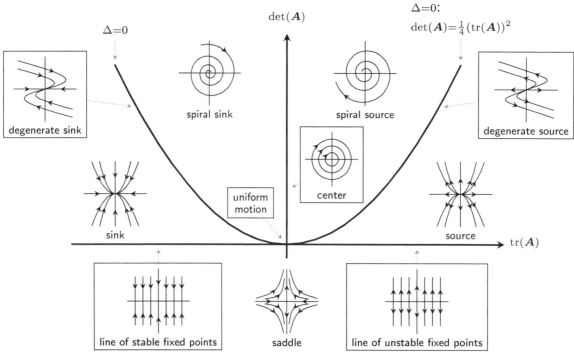

Figure 21. Poincaré diagram for 2x2 matrices: Classification of phase portraits in the $(\det(A), \text{tr}(A))$ plane. Credit to Gernot Salzer, No rights reserved.

1.8 ■ Feedback revisited

In the next step we will use the introduced mathematical concepts and analyze the systemic properties of some network motifs discussed earlier. We will thereby see how and when these motifs give rise to characteristic non-linear dynamic phenomena.

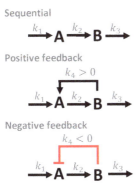

Figure 22. Network structures of the feedback examples as analyzed in Example 9.

Example 9: Feedback

In this example we will perform a detailed, mathematical analysis of the effects of negative and positive feedback in a simple metabolic pathway (Figure 22). We start with the general stoichiometric matrix:

$$N = \begin{bmatrix} v_1 & v_2 & v_3 & v_4 \\ 1 & -1 & 0 & 1 \\ 0 & 1 & -1 & 0 \end{bmatrix} \begin{matrix} A \\ B \end{matrix}$$

We have 2 equations ($m=2$) and 4 reactions ($r=4$). The rank is 2 ($\mathrm{rk}(N)=2$). This means, we can set 2 unknowns k to arbitrary numbers to find 1 of the many possible solutions. We decide to take the influx k_1 as control u and to set k_4 to different values:

$k_4 < 0$: negative feedback
$k_4 = 0$: sequential (no feedback)
$k_4 > 0$: positive feedback

We use mass action kinetics to set up an ODE sys-

tem for the concentrations of A and B:

$$\underbrace{\begin{pmatrix} \dot{A} \\ \dot{B} \end{pmatrix}}_{\dot{s}} = \underbrace{\begin{pmatrix} 1 & -1 & 0 & 1 \\ 0 & 1 & -1 & 0 \end{pmatrix}}_{N} \underbrace{\begin{pmatrix} u \\ k_2 A \\ k_3 B \\ k_4 B \end{pmatrix}}_{v}$$

$$= \begin{pmatrix} -1 & 0 & 1 \\ 1 & -1 & 0 \end{pmatrix} \begin{pmatrix} k_2 A \\ k_3 B \\ k_4 B \end{pmatrix} + \begin{pmatrix} 1 \\ 0 \end{pmatrix} u$$

$$= \begin{pmatrix} -k_2 & 0 & k_4 \\ k_2 & -k_3 & 0 \end{pmatrix} \begin{pmatrix} A \\ B \\ B \end{pmatrix} + \begin{pmatrix} 1 \\ 0 \end{pmatrix} u$$

$$\underbrace{\begin{pmatrix} \dot{A} \\ \dot{B} \end{pmatrix}}_{\dot{x}} = \underbrace{\begin{pmatrix} -k_2 & k_4 \\ k_2 & -k_3 \end{pmatrix}}_{A} \underbrace{\begin{pmatrix} A \\ B \end{pmatrix}}_{x} + \underbrace{\begin{pmatrix} 1 \\ 0 \end{pmatrix}}_{u} u$$

which is the ODE system:

$$\dot{A} = -k_2 A + k_4 B + u$$
$$\dot{B} = k_2 A - k_3 B$$

Let's interpret these equations: We have an influx u, which positively influences the change of A. A part of the substance A turns into B with reaction rate k_2. If term $k_2 A$ is subtracted from the first equation, it negatively influences the change of A. The same term appears in the equation for the change of B. This ensures that no mass is lost. Thus, the term $k_2 A$ positively influences the change of B as it negatively influences the change of A, which makes sense because we want to turn substance A into substance B. The substance B leaves the system with rate k_3. One can see that the loss of substance of B increases with the abundance of B. The feedback via $k_4 B$ influences A depending on the abundance of B. This interaction is modeled as signaling flow. B "only" transmits a signal but is not converted. This means that it does not preserve the mass conservation at this point and it does not reduce B by $k_4 B$.

This is a fundamental and important difference:

- Preserve mass conservation → substrate flow

- Not preserve mass conservation → information flow.

We will now look at the stability of system matrix

A via the Eigenvalues:

$$\left| \begin{pmatrix} -k_2 & k_4 \\ k_2 & -k_3 \end{pmatrix} - \begin{pmatrix} s & 0 \\ 0 & s \end{pmatrix} \right| = \left| \begin{matrix} -(k_2+s) & k_4 \\ k_2 & -(k_3+s) \end{matrix} \right| = 0$$

$$(k_2+s)(k_3+s) - k_2 k_4 = 0$$

$$s^2 + (k_2+k_3)s + k_2(k_3-k_4) = 0$$

We study the case $k_2 = k_3 = 1$:

$$s^2 + 2s + (1-k_4) = 0$$

The related p-q equation is:

$$s_{1,2} = -1 \pm \sqrt{1-(1-k_4)} = -1 \pm \sqrt{k_4}$$

Sequential setting $k_4 = 0$:
Both Eigenvalues are negative
$\Rightarrow s_1 = -1; s_2 = -1$ stable (Type 0)
Positive feedback ($k_4 = 2$):
$\Rightarrow S_1 = 0.4142; S_2 = -2.41$ unstable saddle
Positive feedback ($k_4 = 1$):
$\Rightarrow S_1 = 0; S_2 = -2$ metastable (Type 1)
Negative feedback ($k_4 = -1$):
$\Rightarrow S_{1,2} = -1 \pm i$
Damped oscillations, stable (Type 0)

The following video show the simulation of the dynamic behavior of this pathway with different strengths of negative or positive feedback:
YouTube: Example simulated

1.9 ■ Bifurcation: Split it baby!

Adding to Example 9, we will now tackle the mathematical analysis of another example model of feedback. We will, in addition, analyze how the obtained steady states and dynamic properties depend on the value of a system parameter. Simple feedback can also be modeled as:

$$\dot{x} = -k_{\text{neg}} \cdot x \qquad k_{\text{neg}} > 0, \ x(t=0) = 1$$

which turns out to be a stable steady state, as shown in Figure 16.
The time-course is:

$$x(t) = x_0 \cdot e^{-k_{\text{neg}} \cdot t}$$

as shown in Figure 23 left. Positive feedback, however, might destabilize a system. Let's have a look at combined negative and positive feedback

$$\dot{x} = -k_{\text{neg}} \cdot x + k_{\text{pos}} \cdot x \qquad k_{\text{neg}}, k_{\text{pos}} > 0; \ x(0) = 1$$

whereby the variable x is the **cause** and the change $\dot{x}$ is the **consequence** within the feedback loop. The related solution of this differential equation is:

$$x(t) = x_0 \cdot e^{(k_{\text{pos}} - k_{\text{neg}}) \cdot t}$$

As you see in Figure 23 right below, the system becomes unstable if the parameter exceeds certain values. However, healthy biological systems are never globally unstable, because they run into saturation. Malignancies like cancer and virus production, however, can be seen as biological systems that might lead to global system failure, if outbalancing feedback (medicament, immune system) are too weak to control the disease.

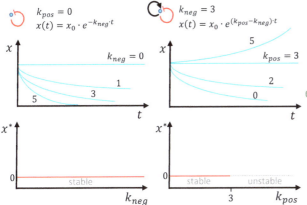

Figure 23. Feedback: On the left-hand side the behavior of simple negative feedback is shown. On the right-hand side the behavior of a system with fixed negative feedback $k_{\text{neg}} = 3$ is shown with different strengths of positive feedback.

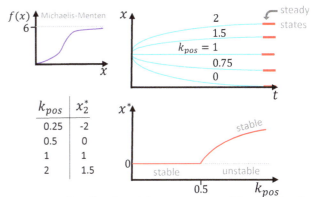

k_{pos}	x_2^*
0.25	-2
0.5	0
1	1
2	1.5

Figure 24. Feedback with saturation: The Michaelis-Menten equation stabilizes the system, so that we get a stable state for each k_{pos}. Compare with Figure 23.

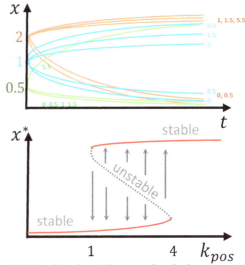

Figure 25. Feedback with complex behavior: Here we get several stable states, but also hysteresis. Compare with Figure 23 and 24.

We can use the Michaelis-Menten function to mimic a saturation for the positive feedback:

$$\dot{x} = -k_{neg} \cdot x + 6 \frac{k_{pos} \cdot x}{k_{pos} \cdot x + 1}$$

$$= -k_{neg} \cdot x + \frac{\overbrace{6}^{v_{max}} \cdot x}{\underbrace{x + 1/k_{pos}}_{K_m}}$$

The solution for the steady state(s) is:

$$-k_{neg} \cdot k_{pos}(x^*)^2 - k_{neg} \cdot x^* + 6k_{pos} \cdot x^* = 0$$
$$x^*(-k_{neg} \cdot k_{pos} \cdot x^* - k_{neg} + 6k_{pos}) = 0$$

Steady states:

Case 1) $\boxed{x_1^* = 0}$

Case 2) $-k_{neg}k_{pos}x^* + 6k_{pos} - k_{neg} = 0$

$$\Rightarrow \boxed{x_2^* = \frac{6k_{pos} - k_{neg}}{k_{pos} \cdot k_{neg}}}$$

For $k_{neg} = 3$: $x_2^* = \frac{6k_{pos} - 3}{3k_{pos}} = \frac{2k_{pos} - 1}{k_{pos}} = 2 - \frac{1}{k_{pos}}$

Stability: Linearization

$$\dot{x} = -k_{neg} \cdot x + 6 \cdot \frac{k_{pos} \cdot x}{k_{pos} \cdot x + 1}$$

$$\Rightarrow \Delta\dot{x} = -k_{neg} \cdot \Delta x + 6\frac{k_{pos}(k_{pos}x^* + 1) - k_{pos}(k_{pos}X^*)}{(k_{pos}x^* - 1)^2}\Delta x$$

Which roughly gives the steady state $x_1^* = 0$

$$\Delta\dot{x} = -k_{neg}\Delta x + 6k_{pos}\Delta x = (6k_{pos} - k_{neg})\Delta x$$

For $k_{neg} = 3$:

$$\Delta\dot{x} = (6k_{pos} - 3)\Delta x$$
$$< 0 \quad \text{for } k_{pos} < 1/2$$
$$> 0 \quad \text{for } k_{pos} > 1/2$$

Stability of this steady state?

$$s = 6k_{pos} - 3$$
$$x_1^* \quad \text{unstable for} \quad k_{pos} > 1/2$$
$$\text{stable for} \quad k_{pos} < 1/2$$

For an illustration of how the steady state(s) and their stability depends on parameter value k_{pos} (bifurcation plot), see Figure 24. The system is not simply unstable beyond a parameter threshold. It changes to the closest stable steady state.

When combining negative feedback and positive feedback with saturation and some basal activation, a bistable system can be obtained (Figure 25). Here, which steady state is reached depends on the parameter value k_{pos} as well as on the initial conditions:

$$\dot{x} = \underbrace{0.2}_{\text{basal level}} - k_{neg}x + 6 \cdot \underbrace{\frac{(k_{pos}x)^5}{(k_{pos}x)^5 + 1}}_{f_{pos}(x)} \quad x(t = 0) = x_0 = 1$$

Summary of key points here:

- Two stable steady states $\Rightarrow$ bistability

- Increase of steady state with k_{pos} not continuous anymore, but step-like. (See next section on bifurcation for more details.)

Neg. feedback $\rightarrow$ stabilizing
$\rightarrow$ oscillating (not shown here)
Pos. feedback $\rightarrow$ (often) destabilizing
$\rightarrow$ bifurcations
$\rightarrow$ bistability, switching behavior
$\rightarrow$ hysteresis

- Contains no information about how long the system needs to reach steady states

- Hysteresis: influence of history. Consequence persists after cause vanishes.

The plot at the bottom of Figure 23 shows the change of a system's steady state depending on parameter value. In the bottom right part of the figure, we see that the system suddenly changes to an unstable steady state, although the previous behavior was approaching an equilibrium. This unstable steady state can lead to different directions—the trajectories split up. The trajectories could split up several times until we, with certain parameter sets, reach chaotic system behavior (not shown).

Definition 6. Bifurcation: Qualitative change of system behavior with parameter displacement.

This is relevant if:

- parameter unknown (welcome to Systems Biology)

- parameter is influenceable (drug target?)

A double negative feedback loop, the toggle switch, is a common motif in differentiation biology, as shown in Figure 26. Waddington's[3] epigenetic landscape is a visualization of the bifurcation valleys and laid a cornerstone of systems biology, epigenetics, and developmental biology. A change of parameters changes the landscape shape. Look back also to Figures 3 and 4 where a dotted line marks the unstable regions. The individual trajectories might go to the adjacent stable stem cell states. If you are as excited and curious as we are, read this paper about bistability, bifurcations, and Waddington's epigenetic landscape [12].

Mathematically, a local bifurcation $(\mathbf{x}_0, \lambda_0)$ appears to be a continuous dynamic system:

$$\dot{\mathbf{x}} = f(\mathbf{x}, \lambda)$$

is linearized around a certain fixed point, and the obtained Jacobian reveals an Eigenvalue with a zero real

---

[3] British biologist and philosopher Conrad Hal Waddington (1905—1975).

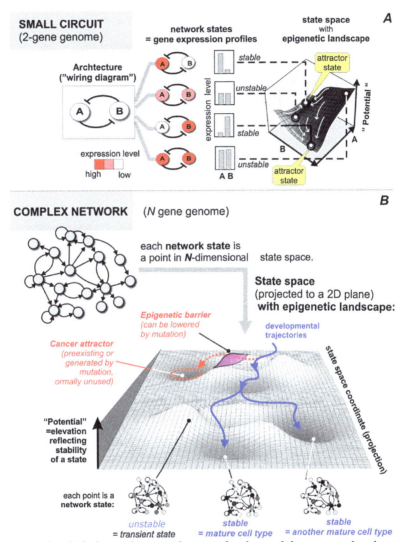

Figure 26. Bifurcation example: (A) depending on the initial values of the protein level states in the toggle switch motif, we can have a stable or unstable situation. An unstable state can fall into the one or the other direction. (B) Developmental trajectories follow the imaginary epigenetic landscape of Conrad Waddington. Adapted from [13]. Copyright © 2009, Elsevier Ltd.

part. If you get stuck here, do not worry. Come back later after you have worked through the rest of the document. The Poincaré–Andronov–Hopf bifurcation,[456] or (in short) Hopf bifurcation, is an interesting case because, beyond this point, the solution becomes oscillatory (complex Eigenvalues). Biological examples are the Lotka–Volterra model for the interplay of predators and prey, the Hodgkin–Huxley model of neuro-electrophysiology, and the Selkov model of glycolysis [14].

Also check out the time profiles and bifurcations as occurring in a cancer-signaling network (Figure 27).

---

[4] Austrian-Hungarian mathematician and astronomer Eberhard Frederich Ferdinand Hopf (1902—1983).

[5] French mathematician, theoretical physicist, engineer, and philosopher Henri Poincaré (1854—1912).

[6] Soviet physicist Aleksandr Aleksandrovich Andronov (1901—1952).

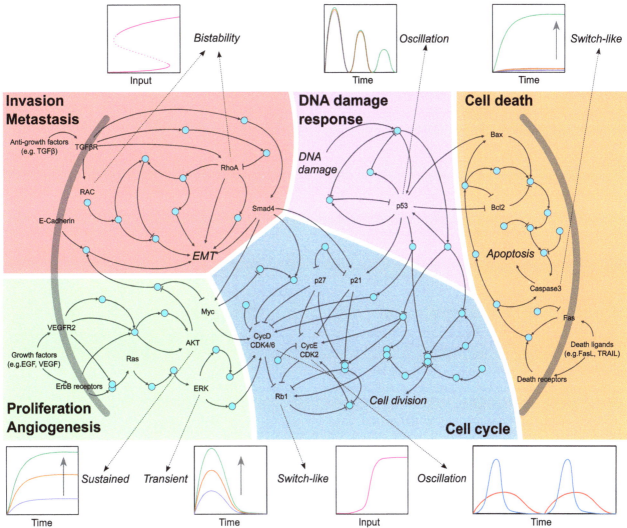

Figure 27. Time profiles in cancer: "Intracellular signaling regulating cancer associated phenotypes and dynamic properties of key signaling proteins. The main signaling cascades regulating cancer hallmarks focused on in this review are depicted. Key signaling proteins in each signaling pathway have various dynamic properties. Cell cycle drivers such as CycD/CDK4 and CycE/CDK2 show oscillatory dynamics, whereas Rb, which is involved in entry for S phase, shows switch-like response. Caspase 3, which mediates programmed cell death (also known as apoptosis), also shows switch-like activation. The tumor suppressor gene p53 displays various dynamics, *e.g.*, oscillation, in response to DNA damage [...]. Rac1 and RhoA, which are involved in epithelial–mesenchymal transition (EMT) and cell motility show bistability. Akt and ERK, which play a key role in multiple cellular processes such as proliferation, differentiation and cell death display sustained and transient dynamics, respectively". Direct quote and source: [15]. Copyright © 2017, Elsevier Ltd.

1.10 ■ Simulation-based analysis of motifs

As mentioned earlier in Example 1 to 6, computer simulation is the method of choice to analyze large or complex dynamical systems. For such systems, analytical solutions cannot usually be obtained anymore. For the simulation within a computational framework, the set of balance equations (ODEs) has to be supplied in a specific form. These equations will then be numerically integrated and the time-course behavior for a given set of initial conditions and parameters will be approximated. By repeatedly simulating a system with different initial conditions and parameters, an overview on the overall system behavior can be obtained. We will not introduce more advanced methods like Parameter Identification or Sensitivity Analysis here (further reading *e.g.* [16]), but will illustrate the basic simulation approach with two examples:

```
********** MODEL NAME
Sniffer

********** MODEL STATES
d/dt(P) = k1p*S - k1d*P
d/dt(R) = k2p*S - k2d*R - k2dp*R*P

P(0) = 0
R(0) = 0

********** MODEL PARAMETERS
S = 0
k1p = 1
k1d = 1
k2p = 1
k2d = 0
k2dp = 1
```

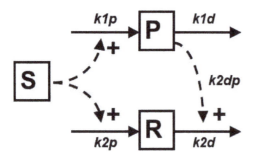

Figure 28. Network structure of the Sniffer motif. Stimulus S activates the production of proteins P and R which both also get degraded. The degradation of P also depends on P. The respective parameter names are given in *italics*.

Example 10: Change detection & adaptation

The Sniffer motif as depicted in Figure 28 is a small motif consisting of two proteins and an activating input stimulus [17]. The respective balance equations are incorporated in the computer model which is given below in the IQM toolbox format [5]. Other tools use similar but slightly different formats. This model allows to simulate the system behavior, *i.e.* the behavior of the output as a function of the input (Figure 29). From this we can see that the response shows perfect adaptation to varying inputs, signaling that there is change but then resettling at the same steady state and being ready to newly respond to another change. Understanding this function of the motif is on a different level compared to just drawing the network interactions. For further reading we recommend the full paper that this extract comes from [17].

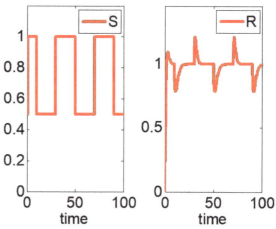

Figure 29. Dynamics of the Sniffer motif. The time course of output R is depicted (right) as a consequence of the varying input S (left).

Example 11: Toggle switch

The toggle switch motif is depicted in Figure 26 and consists of two proteins which are mutually repressing each other's activity, resulting in a double-negative feedback loop. This motif was nicely analyzed by Huang et al. [18] and we recommend reading the paper. One of the variants of the motifs is modeled with the following ODEs:

$$\dot{x}_1 = a_1 \cdot \frac{x_1^n}{tha_1^n + x_1^n} + b_1 \cdot \frac{thb_1^n}{thb_1^n + x_2^n} - k_1 \cdot x_1$$

$$\dot{x}_2 = a_2 \cdot \frac{x_2^n}{tha_2^n + x_2^n} + b_2 \cdot \frac{thb_2^n}{thb_2^n + x_2^n} - k_2 \cdot x_2$$

With the MATLAB code below, using this model and analyzing (with repeated simulations) the effect of the degradation parameter value $k = k1 = k2$ on the number and exact value of the steady states, one obtains the bifurcation plot as shown in Figure 30. The system is bistable for parameter values $k < 1.8$. This results in a phase plot with two stable steady states and one unstable steady state (Figure 31). The slopefield indicates the regions of attraction towards one or the other stable steady state. For parameter values $k > 1.8$ the bistability disappears and the system has only one stable steady state (Figure 32). The respective phase plot shows only that steady state with all the slopefields pointing towards it. This motif plays an important role in many developmental processes (see Figure 26 and [18]).

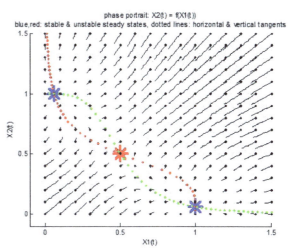

Figure 31. Toggle switch motif. Phase plot for the toggle switch motif (Example 11) with slopefield. For parameter values $k = k1 = k2 = 1$ the system is showing a bistability. Stable steady states are plotted in blue and unstable in red.

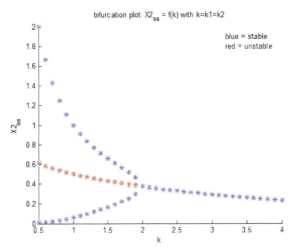

Figure 30. Toggle switch motif. Bifurcation plot for the toggle switch motif (Example 11). The dependency of the steady state(s) for $X2$ as a function of parameter $k = k1 = k2$ is depicted. Stable steady states are plotted in blue and unstable in red.

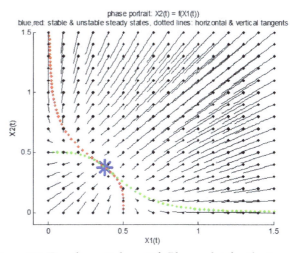

Figure 32. Toggle switch motif. Phase plot for the toggle switch motif (Example 11) with slopefield. For parameter values $k = k1 = k2 = 2$ the system is having only one (stable) steady state (indicated in blue).

```matlab
1  %% Toggle switch motif
2  %Thomas Sauter, University of Luxembourg
3  clear all
4  syms X1 X2
5  syms a1 a2 tha1 tha2 n b1 b2 thb1 thb2 k1 k2
6  eqX1_gen=a1*X1^n/(tha1^n+X1^n) + ...
       b1*thb1^n/(thb1^n+X2^n) - k1*X1
7  eqX2_gen=a2*X2^n/(tha2^n+X2^n) + ...
       b2*thb2^n/(thb2^n+X1^n) - k2*X2
8  % parameters for model A
9  n=4
10 k1=1 %1 %2
11 k2=1 %1 %2
12 tha1=0.5
13 tha2=0.5
14 thb1=0.5
```

```matlab
15  thb2=0.5
16  a1=0 %2de: 1
17  a2=0 %2de: 1
18  b1=1 %4a: 0.5
19  b2=1 %4a: 0.5
20  eqX1=eval(eqX1_gen)
21  eqX2=eval(eqX2_gen)
22  pretty(eqX1)
23  pretty(eqX2)
24  %% step 3
25  [X1_ss_sym,X2_ss_sym]=vpasolve(eqX1,eqX2,X1,X2);
26  X1_ss_tmp=X1_ss_sym; %eval(X1_ss_sym)
27  X2_ss_tmp=X2_ss_sym; %eval(X2_ss_sym)
28  X1_ss=[]; X2_ss=[];
29  for k=1:numel(X1_ss_tmp) %remove complex roots
30      if (imag(X1_ss_tmp(k))==0) && ...
            (imag(X2_ss_tmp(k))==0)
31          X1_ss=[X1_ss; X1_ss_tmp(k)];
32          X2_ss=[X2_ss; X2_ss_tmp(k)];
33      end
34  end
35  X1_ss
36  X2_ss
37  %% step 4 using a for loop
38  syms X1 X2
39  %general linearization
40  Lin = jacobian([eqX1; eqX2],[X1 X2])
41  %around steady states: using for loop
42  for i=1:size(X1_ss)
43      X1=X1_ss(i)
44      X2=X2_ss(i)
45      eval([['Lin' num2str(i)] '= eval(Lin);'])
46      eval([['eig' num2str(i)] '= ...
            eig(eval(Lin))'])
47  end
48  %% ht and vt tangents
49  syms X1 X2
50  [X1_ht,X2_ht,params_ht,conditions_ht] ...
51  =solve(eqX2,X1,X2,'MaxDegree',5,...
52  'ReturnConditions',true)
53  [X1_vt,X2_vt,params_vt,conditions_vt] ...
54  =solve(eqX1,X1,X2,'MaxDegree',5,...
55  'ReturnConditions',true)
56  %% plotting the phase portrait + numerical ...
        simulation of vector field
57  figure; hold on
58  axis([-0.1 1.5 -0.1 1.5]) %Plot range
59  X1_range=linspace(0,1.5,50); %Evaluate all ...
        results in the same range
60  X2_range=linspace(0,1.5,50);
61  % plot steady states
62  for k=1:numel(X1_ss)
63      ss_plot = plot(X1_ss(k),X2_ss(k),'*',...
64      'MarkerSize',20,'LineWidth',4);
65      eig_check=['eig' num2str(k)]; %check ...
            stability
66      if max(real(eval(eig_check)))≥0
67          set(ss_plot,'Color','r') %red for ...
                unstable (≥0)
68      else
69          set(ss_plot,'Color','b') %blue for ...
                stable (<0)
70      end
71  end
72  % plot vertical and horizontal tangents
73  for i=1:size(X1_vt)
74      for k=1:numel(X2_range)
75          z=X2_range(k);
76          %          for l=1:numel(X2_range)
77          %              X2=z;
78          if (imag(eval(X1_vt(i)))==0) && ...
                (imag(eval(X2_vt(i)))==0)
79              plot(eval(X1_vt(i)),eval(X2_vt...
80              (i)),'r.','MarkerSize',10)
81          %          end
82      end
83  end
84  end
85  for i=1:size(X1_ht)
86      for k=1:numel(X2_range)
87          z=X2_range(k);
88          %          for l=1:numel(X2_range)
89          %              X2=X2_range(l);
90          if (imag(eval(X1_ht(i)))==0) && ...
                (imag(eval(X2_ht(i)))==0)
91              plot(eval(X1_ht(i)),eval(X2_ht(i))...
92              ,'g.','MarkerSize',10)
93          end
94      end
95  end
96
97  % numerical computation of example ...
        trajectories
98  X1_ic=[0:0.1:1.5]' %initial conditions on ...
        the x-axis (X1)
99  X2_ic=[0:0.1:1.5]' %initial conditions on ...
        the y-axis (X2)
100
101 syms X1 X2
102 eqX1=subs(eqX1,X1,str2sym('x(1)'));
103 eqX1=subs(eqX1,X2,str2sym('x(2)'))
104 eqX2=subs(eqX2,X1,str2sym('x(1)'));
105 eqX2=subs(eqX2,X2,str2sym('x(2)'))
106 eval(['dxdt=@(t,x,tmp) ([' char(eqX1) ';' ...
        char(eqX2) '])'])
107 t_int=[0 .1] %vectorfield tend=0.1 %full ...
        trajectories tend=1
108 options=[]; % =tmp
109 for i=1:size(X1_ic)
110     for k=1:size(X2_ic)
111         [t, x] = ...
                ode45(dxdt,t_int,[X1_ic(i); ...
                X2_ic(k)],options);
112         plot(x(:,1),x(:,2),'k')
113         plot(X1_ic(i),X2_ic(k),'k.',...
114         'MarkerSize',5)
115     end
116 end
117 hold off
118 title({'phase portrait: X2(t) = ...
        f(X1(t))','blue,red: ...
119 stable & unstable steady states, dotted ...
        lines: ...
120 horizontal & vertical tangents'})
121 xlabel('X1(t)')
122 ylabel('X2(t)')
```

1.11 ■ Additional reading: Cybernetics—The art of creating equilibrium in a world of constraints and possibilities

Cybernetics or control engineering enables us to understand systems and to design the optimal control. The optimal control brings and keeps the system in a favorable state. We can assume that the evolutionary pressure forced living systems to develop elegant control systems.

Let's start with the general state-space representation with a multi-input multi-output (MIMO) system with the block diagram in Figure 33. We are not only interested in the actual system $\dot{x} = Ax$ itself, but also in how to control and how to observe our system. Often we do not know the real state values of our system. We have to measure the states and get an observed output y, such as the pERK level change in a cell culture experiment. Measurements, *e.g.* with western blot, can bias the real data of an untouched system. The observation matrix C describes which real states are pooled to our output and how the experimental analysis modifies the information of the true state values. Besides the analysis, we also know that it is very hard to precisely perform an experiment *e.g.* in cell culture. The pipette might have not been calibrated for a while and who knows what the clumsy intern from the computational department did with it. Thus, all the bias related to the control u is accounted for in the input matrix B. Additionally, not everything that we put into the cell culture (fetal calf serum, DMSO, stains) influences our system $\dot{x} = Ax$, but might directly influence the observed output. Such bias is considered with the feedthrough matrix D. After we have explained the principle ideas behind the state-space representation, we can look at the equations for a simpler single-input single-output (SISO) system of Order Two:

linear system   **non-linear system**

with the matrices and vectors:

$$A = \begin{pmatrix} a_{11} & a_{12} \\ a_{21} & a_{22} \end{pmatrix}; \quad b = \begin{pmatrix} b_1 \\ b_2 \end{pmatrix}; \quad c = \begin{pmatrix} c_1 \\ c_2 \end{pmatrix}$$

$x(\cdot)$ is called the "state vector", $x(t) \in \mathbb{R}^n$;

$y(\cdot)$ is called the "output vector", $y(t) \in \mathbb{R}^q$;

$u(\cdot)$ is called the "input (or control) vector", $u(t) \in \mathbb{R}^p$;

$A(\cdot)$ is the "state (or system) matrix", $\dim[A(\cdot)] = n \times n$,

$B(\cdot)$ is the "input matrix", $\dim[B(\cdot)] = n \times p$,

$C(\cdot)$ is the "output matrix", $\dim[C(\cdot)] = q \times n$,

$D(\cdot)$ is the "feedthrough (or feedforward) matrix" (in cases where the system model does not have a direct feedthrough, $D(\cdot)$ is the zero matrix), $\dim[D(\cdot)] = q \times p$,

and the vector function:

$$\mathbf{f}(\boldsymbol{x}, u) = \begin{pmatrix} f_1(\boldsymbol{x}, u) \\ f_2(\boldsymbol{x}, u) \end{pmatrix}$$

YouTube: State-space representation (all relevant)

Systems often show oscillations and subsiding behaviors over time as shown in Figure 19. The calculation is very difficult as we have to deal with complex integral convolutions. As discussed in linear algebra, we might facilitate our problem by changing the coordinate system. The transfer to a frequency-based coordinate system is indeed a very good idea and has become the main tool of cybernetics. The transfer from the time domain to the frequency domain allows us to treat control problems, for the most part, with simple algebraic operations. The **Laplace transform**[7] splits our function into harmonic sinusoids and exponential decay terms, and is an extension of the **Fourier transform**,[8] which dissects periodic functions into harmonic sinusoids only, as animated in Figure 34. The sinusoid can either occur with different phase shifts or in combination with the cosines. The Laplace transform transfers our mathematical problem from the time-dependent space into the frequency-dependent space (Figure 35) with the complex variable $s = \sigma + i\omega$, as shown in Figure 36. The easily obtained solution in the frequency domain can be transformed back into the time domain by the inverse Laplace transform, as shown in Figure 35. The discrete time equivalent of the Laplace transform is the **Z-transform**.

Some other important concepts in control theory are:

Definition 7. Controllability describes the ability of an external input u to shift the internal state of a system from any initial state x_0 to any other final state x in a finite time interval.

Definition 8. Observability is a measure of how well internal states of a system can be inferred from knowledge of its external outputs.

---

[7] French astronomer and mathematician Pierre-Simon Laplace (1749—1827).
[8] French mathematician and physicist Jean-Baptiste Joseph Fourier (1768—1830).

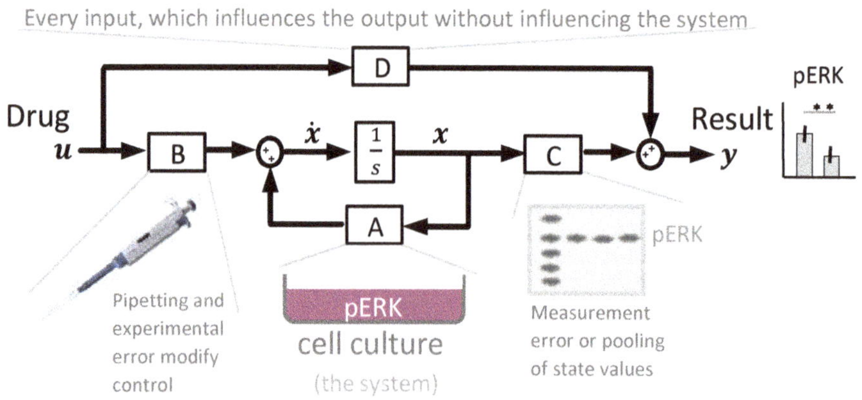

Figure 33. Block diagram of a multi-input multi-output (MIMO) system with associations to experimental cell biology. The operation $\frac{1}{s}$ is an integral in the frequency space (Laplace transform) that turns the state derivatives into state values. See text for explanation.

Definition 9. Transfer function: A transfer function is a mathematical function which gives for each possible input value the corresponding output value.

Two types of principal design can be distinguished: **open loop** and a **closed loop**. An open loop design requires a complete understanding of all physics involved. An open loop system approaches the desired state after a certain time, but might not exactly reach it due to disturbances. A closed system compares the system output with a set point and permanently tries to reduce the error. A closed system reaches the set point despite disturbances and can reach the desired state faster with oscillations around the target. A good control system brings our system into the desired state in a short time with high precision despite possible disturbances. Watch the first five videos on control theory here:
YouTube: Control theory lectures (videos 1-5).

Time domain

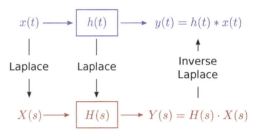

Frequency domain

Figure 34. Fourier transform: a periodic function in the time domain (f, red, magnitude over time) can be split into several sinus and cosinus functions or several sinus functions with phase shift. These are peaks in the frequency domain ($\hat{f}$, blue, amplitude over frequency). Credit to Lucas V Barbosa, Wikimedia, Licence: Public Domain Dedication.

Figure 35. Time and frequency domain. Input $x(t)$, impulse-response $h(t)$, and output $y(t)$ in the time domain, and their analogues $(X(s), H(s), Y(s))$ in the frequency domain. We have complicated convolution operations ($*$) in the time domain, and we have simple algebraic operations ($\cdot$) in the frequency domain. Picture source: Wikimedia, Licence: CC0 1.0 Universal Public Domain Dedication.

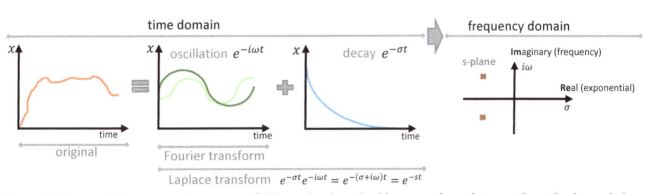

Figure 36. Time and frequency domain: an ODE can be described by a number of sinusoids with phase shift or sinusoids and cosinoids. The process is known as the Fourier transform. This is not enough for control engineering, but if we represent the original ODE as a sum of sinusoids and exponential decay functions, it is sufficient. This process is called the Laplace transform, which allow us to work in the s-plane and frequency domain.

2. Basics of mathematics

Before you go through the following chapter, consider watching this series of YouTube videos on calculus (3 h):

YouTube: Essence of calculus

Solving quadratic equations

p-q formula

The solution of:

$$x^2 + px + q = 0$$

is:

$$x_{1,2} = -\frac{p}{2} \pm \sqrt{\left(\frac{p}{2}\right)^2 - q}.$$

a-b-c formula

The solution of

$$ax^2 + bx + c = 0$$

is:

$$x_{1,2} = \frac{-b \pm \sqrt{b^2 - 4ac}}{2a}$$

Complex numbers

Scientists had huge difficulties calculating the roots of negative numbers for a long time. The situation improved after defining i as the solution of the equation $x^2 = -1$ resulting in $i^2 = -1$ and $i = \sqrt{-1}$. We get a complex number:

$$z = a + bi$$

with a real part a and an imaginary part b. This is needed in this course, when you apply the p-q-equation. Moreover, it is important to know that the imaginary part is directly linked to the sine function via Euler's formula $e^{i\phi} = \cos(\phi) + i \cdot \sin(\phi)$:

$$z = re^{i\phi}$$
$$= r(\cos(\phi) + i \cdot \sin(\phi))$$

$$e^{st} = e^{[Re(s)t + iIm(s)t]}$$

$$= e^{Re(s)t} \cdot e^{iIm(s)t}$$

$$= e^{Re(s)t}[\cos(Im(s)t) + i \cdot \sin(Im(s)t)]$$

Watch Part 1 to 10 of this series on imaginary numbers to get a feeling:

YouTube: Series on imaginary numbers

Derivative

The derivative is defined as:

$$f'(a) = \lim_{h \to 0} \frac{f(a+h) - f(a)}{h}.$$

where the distance h becomes infinitesimally small. The derivative symbolizes the tangent at a function, as shown in Figure 37.

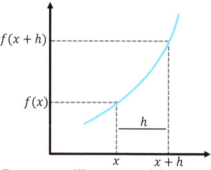

Figure 37. Derivative. Illustration of the derivative of a function is obtained.

Differentiation

$$y = x^n$$

$$\frac{dy}{dx} = y' = n \cdot x^{n-1}$$

Example 1: Differentiation

$$y = x^3 \qquad\qquad y = 4x^4$$
$$\dot{y} = 3x^2 \qquad\qquad \dot{y} = 16x^3$$

Product rule

$$y = u(x) \cdot v(x)$$
$$y' = u'(x) \cdot v(x) + u(x) \cdot v'(x)$$

Example 2: Product rule

$$y = x^2 e^x$$
$$\dot{y} = 2x \cdot e^x + x^2 \cdot e^x$$

Quotient rule

$$y = \frac{u(x)}{v(x)} = \frac{\text{numerator}}{\text{denominator}}$$

$$y' = \frac{u'(x) \cdot v(x) - u(x) \cdot v'(x)}{v(x)^2}$$

Example 3: Quotient rule

$$\frac{d}{dx}\frac{e^x}{x^2} = \frac{\left(\frac{d}{dx}e^x\right)(x^2) - (e^x)\left(\frac{d}{dx}x^2\right)}{(x^2)^2}$$

$$= \frac{(e^x)(x^2) - (e^x)(2x)}{x^4}$$

$$= \frac{e^x(x-2)}{x^3}.$$

(Source: Wikipedia)

Chain rule

$$y = u(v(x))$$

with $y = u(z)$ and $z = v(x)$

$$\frac{dy}{dx} = y' = \frac{dy}{dz} \cdot \frac{dz}{dx} = u'(z)v'(x) = f'(v(x))v'(x).$$

Example 4: Chain rule

$$y = (2x+1)^2$$
$$\dot{y} = 2(2x+1) \cdot 2 = 4(2x+1)$$

■ Basic integrals

An integral is basically the area under the curve (AUC). The term AUC is often used, *e.g.*, in pharmarcokinetics.

$$\int x^n \, dx = \frac{1}{n+1}x^{n+1}, \qquad \text{for } n \neq -1$$

$$\int u \, dv = uv - \int v du$$

$$\int e^x \, dx = e^x$$

$$\int a^x \, dx = \frac{1}{\ln a}a^x$$

$$\int \frac{1}{x} \, dx = \ln|x|$$

$$\int \ln x \, dx = x\ln x - x$$

■ Transform one ODE of nth order into n ODEs of 1st order
Approach
1. Introduce n new variables:

$$x_1 = x$$
$$x_2 = \dot{x}$$
$$x_3 = \ddot{x}$$
$$\vdots$$
$$x_n = \overset{(n)}{x}$$

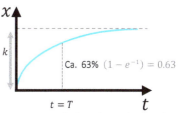

Figure 38. Time course behavior of an inhomogeneous ODE, given as $T\dot{x} + x = u$. The Euler number is $e = 2.718$.

2. Write $(n-1)$ equations of the form:

$$\dot{x}_i = x_{i+1} \qquad ; i = 1\ldots(n-1)$$

3. Transform original ODE into 1 ODE of 1st-order by replacing the derivatives with x_i.

Example 5: Converting an ODE of nth to n 1st order

The equation of 2nd order is:

$$\ddot{x} + 4\dot{x} + 5x = 0 \qquad (n = 2)$$

With the ansatz

$$x \equiv x_1$$
$$\dot{x} \equiv x_2 = \dot{x}_1$$
$$\ddot{x} \equiv x_3 = \dot{x}_2$$

we obtain for our system:

$$\dot{x}_1 = x_2$$
$$\dot{x}_2 = -4x_2 - 5x_1$$

with the matrix form:

$$\begin{pmatrix} \dot{x}_1 \\ \dot{x}_2 \end{pmatrix} = \begin{pmatrix} 0 & 1 \\ -5 & -4 \end{pmatrix} \begin{pmatrix} x_1 \\ x_2 \end{pmatrix}$$

■ Transform two ODEs of 1st order into one ODE of 2nd order
Approach: 1. Solve one ODE for the other state variable:

$$\dot{x}_1 = f(x_1, x_2, u) \rightarrow \text{solve for } x_2$$

2. Differentiate the solution.
3. Insert into other ODEs.

Example 6: Converting ODEs from 1st to 2nd order

We begin with the equation set:

$$\dot{x}_1 = a_{11}x_1 + a_{12}x_2 + b_1 u$$
$$\dot{x}_2 = a_{21}x_1 + a_{22}x_2 + b_2 u$$

The first equation is differentiated with respect to time t:

$$x_2 = \frac{1}{a_{12}}(\dot{x}_1 - a_{11}x_1 - b_1 u)$$
$$\dot{x}_2 = \frac{1}{a_{12}}(\ddot{x}_1 - a_{11}\dot{x}_1 - b_1\dot{u})$$

Transfer of the functions $x_2(x_1, u)$ and $\dot{x}_2(x_1, u)$ to the second equation delivers:

$$\frac{1}{a_{12}}(\ddot{x}_1 - a_{11}\dot{x}_1 - b_1\dot{u}) = a_{21}x_1 + b_2 u$$
$$+ \frac{a_{22}}{a_{12}}(\dot{x}_1 - a_{11}x_1 - b_1 u)$$

which in a nicer form gives:

$$\ddot{x}_1 = (a_{11} + a_{22})\dot{x}_1 - (a_{11}a_{22} - a_{12}a_{21})x_1$$
$$+ b_1\dot{u} + (a_{12}b_2 - a_{22}b_1)u$$

■ **Linearization**

Biological systems are usually non-linear and the measured effects rarely follow the law of additivity (Figure 39). Still, the linearization of equations helps us for the following reasons:

- Linear equations are mathematically easier to solve/treat

- Linear system properties are transferable to non-linear systems around the approximated area

- A unique steady-state exists

- They have a simple controller design

- The response to a variety of simuli can be separated into linear combinations of the system down to individual stimuli.

- Additivity $f(x_1, x_2) = f(x_1) + f(x_2)$ (Figure 39)

- Scalability $c \cdot f(x_1) = f(c \cdot x_1)$ (Figure 39)

For the linearization, we can use the Taylor approximation:

$$f(x) = g(x^*) + \frac{g'(x^*)}{1!}(x - x^*)$$
$$+ \underbrace{\frac{g''(x^*)}{2!}(x - x^*)^2 + \frac{g'''(x^*)}{3!}(x - x^*)^3 + \cdots}_{\text{higher order terms often neglected}}$$

around a point of interest x^*, which can be written as

$$f(x) - g(x^*) = \dot{\Delta}x = g'(x^*)\Delta x$$

whereby the difference between original function $g(x^*)$ and its approximation $f(x^*)$ is emphasized. A set of non-linear equations:

$$\dot{\mathbf{x}} = \mathbf{g}(\mathbf{x}) = \begin{bmatrix} g_1(\mathbf{x}) \\ g_2(\mathbf{x}) \\ \vdots \\ g_n(\mathbf{x}) \end{bmatrix}$$

can be approximated by the general form:

$$\mathbf{f}(\mathbf{x}) = \mathbf{g}(\mathbf{x}^*) + \left. \frac{\partial \mathbf{fg}(\mathbf{x})}{\partial \mathbf{x}} \right|_{x^*} \cdot \Delta \mathbf{x}$$

with the Jacobian matrix:

$$J = \left. \frac{\partial \mathbf{g}(\mathbf{x})}{\partial \mathbf{x}} \right|_{x^*} = \begin{bmatrix} \frac{\partial g_1}{\partial x_1} & \cdots & \frac{\partial g_1}{\partial x_n} \\ \vdots & \ddots & \vdots \\ \frac{\partial g_m}{\partial x_1} & \cdots & \frac{\partial g_m}{\partial x_n} \end{bmatrix}$$

A linearized system in the general form is then:

$$\dot{\mathbf{x}} = J\Delta\mathbf{x}.$$

The Taylor approximation is a very important concept. It delivers a function that, with increasing terms, aligns with any continuous, smooth function around a certain point. However, if one moves too far away from the point of interest, the differences can be extreme.
YouTube: Taylor series explained

Because we are only interested in a linear tangent at a steady state x^*, we neglect all higher order terms and look only at the function value plus the first term of the Taylor approximation:

$$f(x) = g(x^*) + (x - x^*)g'(x^*)$$

of $x = x^* + \Delta x$ with $\Delta x = (x - x^*)$.

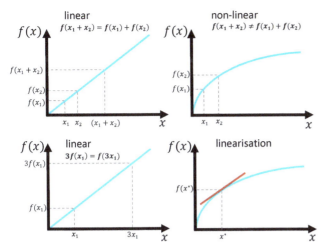

Figure 39. Linear vs non-linear behavior. Linear systems show additivity and scalability.

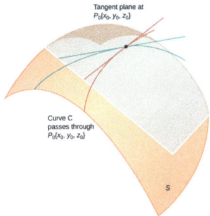

Figure 40. Linearization of a multi-dimensional equation at one point with a plane. Source: LibreTexts Library for mathematics. If the point is an equilibrium point, the plane is horizontal. Source: LibreTexts Library. Licence: CC BY-NC-SA 4.0 (Fair Use).

Example 7: Linearization 1

First, we determine the steady state:

$$g(x) = \dot{x} = 2x - 1$$
$$g(x)' = \ddot{x} = 2$$

$$\dot{x} = 2x^* - 1 = 0 \qquad \Rightarrow x^* = \frac{1}{2}$$

Second, we approximate the function $g(x)$ with a

linear function $f(x)$ at this point:

$$f(x) = \dot{x} = (2x^* - 1) + 2\Delta x$$
$$= \dot{x} = (2\frac{1}{2} - 1) + 2\Delta x = 2\Delta x$$

Example 8: Linearization 2

First, we determine the steady state:

$$g(x) = \dot{x} = x^2 + 2x - 1$$
$$g(x)' = \ddot{x} = 2x + 2$$

$$\dot{x} = (x^*)^2 + 2x^* - 1 \stackrel{!}{=} 0 \Rightarrow$$

$$x^*_{1,2} = \frac{-b \pm \sqrt{b^2 - 4ac}}{2a} = \frac{-2 \pm \sqrt{4 + 4}}{2}$$

$$= -1 \pm \sqrt{2}$$

Second, we approximate the function $g(x)$ with a linear function $f(x)$ at this point:

$$f(x) = \dot{x} = ((x^*)^2 + 2x^* - 1) + (2x^* + 2)\Delta x$$

For $x^*_1 = -1 + \sqrt{2}$:

$$= \dot{x} = ((-1 + \sqrt{2})^2 + 2(-1 + \sqrt{2}) - 1)$$
$$+ (2(-1 + \sqrt{2}) + 2)\Delta x$$
$$= 2\sqrt{2}\Delta x$$

For $x^*_2 = -1 - \sqrt{2}$:

$$= \dot{x} = ((-1 - \sqrt{2})^2 + 2(-1 - \sqrt{2}) - 1)$$
$$+ (2(-1 - \sqrt{2}) + 2)\Delta x$$
$$= -2\sqrt{2}\Delta x$$

The steady state at $x^*_1 = -1 + \sqrt{2}$ is unstable and at $x^*_2 = -1 - \sqrt{2}$ is stable. See Figure 16.

Example 9: Linearization 3

First, we determine the steady state

$$g(x) = \dot{x} = 2x^2 - 2x - 12$$
$$g(x)' = \ddot{x} = 4x - 2$$

$$\dot{x} = 2(x^*)^2 - 2x^* - 12 \stackrel{!}{=} 0$$

$$x^*_{1,2} = \frac{-b \pm \sqrt{b^2 - 4ac}}{2a} = \frac{2 \pm \sqrt{4 + 96}}{4}$$

$$= \frac{1}{2} \pm \frac{5}{2}$$

and obtain the steady states at: $x^*_1 = 3$; $x^*_2 = -2$

Second, we approximate the function $g(x)$ with a linear function $f(x)$ at this point:

$$f(x) = \dot{x} = (2(x^*)^2 - 2x^* - 12) + (4x^* - 2)\Delta x$$
$$= \dot{x} = 0 + (4x^* - 2)\Delta x$$

For $x_1^* = 3$:

$$= 10\Delta x$$

For $x_2^* = -2$:

$$= -10\Delta x$$

The steady state at $x_1^* = 3$ is unstable and at $x_2^* = -2$ is stable. See Figure 16.

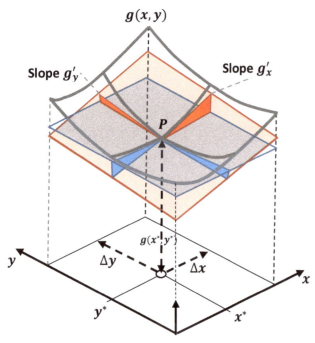

Figure 41. Linearization of a function (hanging carpet) in respect to two directions. The blue plane is horizontal at the steady state of function f. The red plane gives the slopes if shifted slightly moving away from the steady state point P.

Example 10: Linearization 4

If we linearize a system with two non-linear equations:

$$g_1(x,y) = \dot{x} = (1 - x - y)x$$
$$= -x^2 + (1 - y)x$$
$$= -x^2 + x - xy$$
$$g_1(x,y)_x = -2x + 1 - y$$
$$g_1(x,y)_y = -x$$
$$g_2(x,y) = \dot{y} = (1 - x - y)y$$
$$= -y^2 + (1 - x)y$$
$$= -y^2 + y - xy$$
$$g_2(x,y)_x = -y$$
$$g_2(x,y)_y = -2y + 1 - x$$

we approximate it with a plane at a particular point

$$f(x,y) = g(x^*, y^*) + g_x'(x^*, y^*) \cdot \Delta x + g_y'(x^*, y^*) \cdot \Delta y$$

such as shown in Figure 40. But our points of interest are, as usual, steady-state solutions. For these steady-state solutions, the linearized system is:

$$f_1(x,y) = \dot{x} \quad = 0 + (1 - 2x^* - y^*)\Delta x - x^*\Delta y$$
$$f_2(x,y) = \dot{x} \quad = 0 + (1 - 2y^* - x^*)\Delta y - y^*\Delta x$$

Example 11: Linearization 5

We want a linear approximation on a higher-order ODE

$$\ddot{x} + \dot{x} + x = 1$$

on the steady state $x^* = 1$. The equation is rewritten:

$$g(x, \dot{x}, \ddot{x}) = \ddot{x} + \dot{x} + x - 1 = 0,$$

and we treat any derivative as an independent variable. For:

$$f(x, \dot{x}, \ddot{x}) = g(x^*) + g_x'(x^*) \cdot \Delta x + g_{\dot{x}}'(x^*) \cdot \Delta \dot{x} + g_{\ddot{x}}'(x^*) \cdot \Delta \ddot{x}$$

with

$$g_x'(x^*) = 1$$
$$g_{\dot{x}}'(x^*) = 1$$
$$g_{\ddot{x}}'(x^*) = 1$$

we simply get:

$$f(x, \dot{x}, \ddot{x}) = \Delta x + \Delta \dot{x} + \Delta \ddot{x} = 0$$

Example 12: Linearization 6

We aim for a linear approximation of a higher order non-linear ODE:

$$g(x, \dot{x}, \ddot{x}) = \ddot{x} + \dot{x} + x^2 + 5 = 0$$

With

$$f(x, \dot{x}, \ddot{x}) = g(x^*) + g'_x(x^*) \cdot \Delta x + g'_{\dot{x}}(x^*) \cdot \Delta \dot{x} + g'_{\ddot{x}}(x^*) \cdot \Delta \ddot{x}$$

and with

$$g'_x(x^*) = 2x$$
$$g'_{\dot{x}}(x^*) = 1$$
$$g'_{\ddot{x}}(x^*) = 1$$

we get:

$$f(x, \dot{x}, \ddot{x}) = \underbrace{\ddot{x}^* + \dot{x}^* + (x^*)^2 + 5}_{=0} + 2\Delta x + \Delta \dot{x} + \Delta \ddot{x} = 0$$

References

[1] Ertugrul M Ozbudak, Mukund Thattai, Han N Lim, Boris I Shraiman, and Alexander Van Oudenaarden. Multistability in the lactose utilization network of Escherichia coli. *Nature*, 427(6976):737, 2004. https://doi.org/10.1038/nature02298.

[2] Maria Herberg and Ingo Roeder. Computational modelling of embryonic stem-cell fate control. *Development*, 142(13):2250–2260, 2015. https://doi.org/10.1242/dev.116343.

[3] Uri Alon. *An introduction to systems biology: design principles of biological circuits*. CRC Press, 2006. https://doi.org/10.1201/9781420011432.

[4] Uri Alon. Network motifs: Theory and experimental approaches. *Nature Reviews Genetics*, 8(6):450, 2007. https://doi.org/10.1038/nrg2102.

[5] M. Sunnaker and H. Schmidt. IQM Tools - Efficient state of the art modeling across pharmacometrics and systems pharmacology. *Pharmocokin Pharmacodyn*, 43:69–70, 2016.

[6] James E Ferrell Jr. Self-perpetuating states in signal transduction: Positive feedback, double-negative feedback and bistability. *Current Opinion in Cell Biology*, 14(2):140–148, 2002. https://doi.org/10.1016/S0955-0674(02)00314-9.

[7] Sabrina L Spencer and Peter K Sorger. Measuring and modeling apoptosis in single cells. *Cell*, 144(6):926–939, 2011. https://doi.org/10.1016/j.cell.2011.03.002.

[8] Joachim W Schmid, Klaus Mauch, Matthias Reuss, Ernst D Gilles, and Andreas Kremling. Metabolic design based on a coupled gene expression—metabolic network model of tryptophan production in Escherichia coli. *Metabolic Engineering*, 6(4):364–377, 2004. https://doi.org/10.1016/j.ymben.2004.06.003.

[9] Hartmut Bossel. *Modeling and Simulation*. 1994. https://doi.org/10.1201/9781315275574.

[10] James E Ferrell Jr. Perfect and near-perfect adaptation in cell signaling. *Cell Systems*, 2(2):62–67, 2016. https://doi.org/10.1016/j.cels.2016.02.006.

[11] Robin E Ferner and Jeffrey K Aronson. Cato Guldberg and Peter Waage, the history of the law of mass action, and its relevance to clinical pharmacology. *British Journal of Clinical Pharmacology*, 81(1):52–55, 2016. https://doi.org/10.1111/bcp.12721.

[12] James E Ferrell Jr. Bistability, bifurcations, and waddington's epigenetic landscape. *Current Biology*, 22(11):R458–R466, 2012. https://doi.org/10.1016/j.cub.2012.03.045.

[13] Sui Huang, Ingemar Ernberg, and Stuart Kauffman. Cancer attractors: a systems view of tumors from a gene network dynamics and developmental perspective. In *Seminars in Cell & Developmental Biology*, volume 20, pages 869–876. Elsevier, 2009. https://doi.org/10.1016/j.semcdb.2009.07.003.

[14] EE Sel'Kov. Self-oscillations in glycolysis 1. a simple kinetic model. *European Journal of Biochemistry*, 4(1):79–86, 1968. https://doi.org/10.1111/j.1432-1033.1968.tb00175.x.

[15] Shigeyuki Magi, Kazunari Iwamoto, and Mariko Okada-Hatakeyama. Current status of mathematical modeling of cancer–from the viewpoint of cancer hallmarks. *Current Opinion in Systems Biology*, 2:39–48, 2017. https://doi.org/10.1016/j.coisb.2017.02.008.

[16] Edda Klipp, Wolfram Liebermeister, Christoph Wierling, and Axel Kowald. *Systems Biology*. Wiley-VCH Verlag, Weinheim, Germany, 2016.

[17] J. J. Tyson, K. C. Chen, and B. Novak. Sniffers, buzzers, toggles and blinkers: Dynamics of regulatory and signaling pathways in the cell. *Curr Opin Cell Biol*, 15(2):221–231, 2003. https://doi.org/10.1016/S0955-0674(03)00017-6.

[18] S. Huang, Y. P. Guo, G. May, and T. Enver. Bifurcation dynamics in lineage-commitment in bipotent progenitor cells. *Dev Biol*, 305(2):695–713, 2007. https://doi.org/10.1016/j.ydbio.2007.02.036.

3. Exercises

Steady states

Calculate the steady states for systems (3.1) to (3.8):

$$\dot{x} = 2x - 1 \tag{3.1}$$

$$\dot{x} = x^2 + 2x - 1 \tag{3.2}$$

$$\dot{x} = \sin(x) \tag{3.3}$$

$$\ddot{x} + \dot{x} + x = 1 \tag{3.4}$$

$$\ddot{x} + 2\dot{x} + x^2 - 5x + 6 = 0 \tag{3.5}$$

$$\frac{d}{dt}\begin{pmatrix} x_1 \\ x_2 \end{pmatrix} = \begin{pmatrix} 3 & -2 \\ 1 & 5 \end{pmatrix}\begin{pmatrix} x_1 \\ x_2 \end{pmatrix} + \begin{pmatrix} 1 \\ 0 \end{pmatrix} u \tag{3.6}$$

$$\begin{aligned} \dot{x} &= (1 - x - y)x \\ \dot{y} &= (1 - x - y)y \end{aligned} \tag{3.7}$$

$$\begin{aligned} \dot{u} &= a - (b+1)u + u^2 v \\ \dot{v} &= bu - u^2 v \end{aligned} \tag{3.8}$$

Stability

Test the following systems for the stability of their steady state(s). Draw, therefore, $\dot{x}$ over x.

$$\dot{x} = 3x \tag{3.9}$$

$$\dot{x} = -5x \tag{3.10}$$

$$\dot{x} = -x^2 + 0.5 \tag{3.11}$$

$$\dot{x} = -x^3 + \frac{x}{2} \tag{3.12}$$

Eigenvalues and stability of the steady state

Determine the Eigenvalues and the stability of steady states of following tasks:

$$\frac{d}{dt}\begin{pmatrix} x_1 \\ x_2 \end{pmatrix} = \begin{pmatrix} -1 & 4 \\ 0 & -3 \end{pmatrix}\begin{pmatrix} x_1 \\ x_2 \end{pmatrix} \tag{3.13}$$

$$\frac{d}{dt}\begin{pmatrix} x_1 \\ x_2 \end{pmatrix} = \begin{pmatrix} 3 & 8 \\ 1 & 1 \end{pmatrix}\begin{pmatrix} x_1 \\ x_2 \end{pmatrix} \tag{3.14}$$

$$\frac{d}{dt}\begin{pmatrix} x_1 \\ x_2 \end{pmatrix} = \begin{pmatrix} -1 & 3 \\ -1 & 1 \end{pmatrix}\begin{pmatrix} x_1 \\ x_2 \end{pmatrix} \tag{3.15}$$

Population growth

The growth of a specific population is described with the following model:

$$\frac{dX}{dt} = (2 - X)(X - 1)(X - k) \qquad (\text{with } k > 0 \text{ and } X(0) = X_0)$$

a) Derive the steady states of the system.

b) Linearize the differential equation around the steady states.

c) Calculate the Eigenvalues of the linearized system in the steady states and determine thereof the stability of the individual steady states. For which values of k do you get stable steady states?

In the following, $k = 5$ holds:

d) Mark within a X-over-t diagram the domains where $\frac{dX}{dt}$ is positive, then respectively negative.

e) Draw $X(t)$ in a diagram using the results of parts a)-d) for different initial conditions of X.

1st-order system

The growth of a specific population is described with the following model:

$$\frac{dN}{dt} = rN(2 - N)(N - 1) \qquad (\text{with } r > 0 \text{ and } N(0) = N_0)$$

a) Derive the steady states of the system.

b) Linearize the differential equation around the steady states.

c) Calculate the Eigenvalues of the linearized system in the steady states and test for stability of the individual steady states.

d) Mark within a t, N diagram the domains where $\frac{dN}{dt}$ is positive, then respectively negative.

e) Draw $N(t)$ in a diagram using the results of parts a)-d) for different initial conditions of N. Also Calculate the position of the inflection points.

Characteristic equation of a 2nd-order system

The following characteristic equation of a 2nd-order system is given:

$$(s - a + T)(s - 1 + aT) = 0 \quad (a, T \text{ are real})$$

a) What is the corresponding homogeneous differential equation?

b) For which values of the parameters a and T do we have *one single* Eigenvalue on the stability bound?

c) For which values of the parameters a and T is the system stable, metastable, or unstable?

d) Draw the results of c) in a stability diagram with abscissa T and ordinate a.

▦ **Biomass growth in a bioreactor**

A batch fermentation is carried out in an ideally mixed tank reactor. At the beginning, substrate and biomass is provided with concentrations $S_0 = S(t = 0)$ and $X_0 = X(t = 0)$.

The reaction volume of the reactor (V_R) is considered to be constant. The specific growth rate of the microorganisms shall be μ. The yield coefficient[9] shall be Y_{XS}. The biomass is linearly proportional to the decline in substrate $\dot{X} = -Y_{XS}\dot{S}$ and this anti-correlating behavior is thus a measurement of efficiency.

a) Set up balance equations for biomass (X) and substrate (S) concentrations.

b) Derive $S(t)$ as a function of $X(t)$, X_0 und S_0.

c) Derive a differential equation for X which does not depend on the substrate concentration anymore $[\dot{X} = f(X)]$ using the assumption $\mu = cS$.

d) Repeat this while assuming a logistic growth rate $\mu = a - bX$ with $(a, b > 0)$.

e) Compare the differential equations obtained for X in c) and d). What would a and b have to be in order to obtain identical differential equations?

f) Calculate the steady states (X^*, S^*) of system d)

g) Show that the following *ansatz* is a solution for $X(t)$:

$$X(t) = \frac{X^*}{1 + c_1 e^{-c_2 t}} \tag{3.16}$$

How do you have to choose c_1 and c_2 $(c_1, c_2 \neq 0)$?

▦ **Prey-Predator Model 1**

We have two versions of the Prey-Predator Model of Lotka and Volterra. Prey X grows independently of Predator Y and is depleted by interacting with Y. Thereby Predator Y grows and, in turn, is depleted by a natural death rate. The most simple differential equation model reflecting these relations is:

$$\dot{X} = X(1 - Y) \qquad \qquad \text{I}$$
$$\dot{Y} = Y(X - 1) \qquad \qquad \text{II}$$

a) Assign the facts described above to the individual model terms in the system.

b) Calculate the steady states of the system.

c) How can the behavior in vicinity of the steady states be approximated (without calculations)?

[9] "Yield based on substrate (Yx/s) or oxygen consumption (Yx/o) is a very important parameter. This parameter indicates how efficient a fermentation is. At the same time it is very closely related with the maintenance coefficient (m). By means of yield and maintenance coefficient it is possible to estimate the proportion of energy that cells consume in biomass and metabolites synthesis and the proportion of energy that allows the cells to maintain their capability for their biological performance." Direct quote [?].

▦ **Prey-Predator Model 2**

In the following, a more complex Lotka-Voltera model shall be investigated. The growth rate of the prey is included with a parameter $\mu > 0$ and the predator's growth rate is modeled with a yield coefficient $\alpha > 0$. Furthermore an influx of predators into the system is assumed. The resulting model description is:

$$\dot{X} = X(\mu - Y) \qquad \qquad \text{I}$$
$$\dot{Y} = Y(\alpha X - 1) + 1 \qquad \qquad \text{II}$$

a) Assign the facts described above to the individual model terms in this system.

b) Calculate the steady states of system as a function of parameters α and μ.

c) Which condition has to be fulfilled for parameter μ to obtain only biologically reasonable steady states?

d) Draw the steady states for $\mu = 2$ and $\alpha = [1/5; 1/4; 1/3; 1/2; 1]$ in a diagram (x-axis: X, y-axis: Y).

▦ **Prey-Predator Model 3 (fish)**

Small and big fishes are living in a lake. The small fishes (X) depend on plankton which is available in excess. Big fishes (Y) feed on the small ones. If only small fishes were present, it would grow exponentially, *i.e.* their number grows with a constant specific growth rate α $(\alpha > 0)$. In the absence of small fishes, the big fishes would die out with a decay rate β $(\beta > 0)$. The following simple equations describe these facts using the parameters $\alpha, \beta, \gamma, \delta > 0$:

$$\frac{d}{dt}X = (\alpha - \gamma Y)X \qquad \qquad \text{I}$$
$$\frac{d}{dt}Y = -(\beta - \delta X)Y \qquad \qquad \text{II}$$

a) Determine the steady states of this simple ecological model.

b) Linearize the differential equations around the steady states and transform the linearized equations into state-space description (states, inputs, outputs in matrix/vector representation).

c) Determine for every steady state the Eigenvalues of the linearized system and deduce from that the stability of the system in the vicinity of the respective steady state.

▦ **Prey-Predator Model 4 (fish)**

The previously introduced fish population model is modified as follows: it is assumed that the small fishes (X) are furthermore depleted via social attrition effects within their own population. This depletion depends on the probability that two small fishes meet. It is therefore modeled proportional to X^2. The extended population model reads as follows with $\alpha = \beta = \gamma = \delta = 1; \varepsilon = 0.5$:

$$\frac{d}{dt}X = (\alpha - \gamma Y)X - \varepsilon X^2 \qquad \text{I}$$

$$\frac{d}{dt}Y = -(\beta - \delta X)Y \qquad \text{II}$$

a) Determine the steady states of the system.

b) Linearize the differential equations around the steady state and transform them into state-space description (states, inputs, outputs in matrix/vector representation).

c) Calculate the Eigenvalues of the linearized system at the steady states and determine thereof the behavior of the system in the vicinity of the steady states.

d) Determine the equations of the vertical and horizontal tangents.

e) Optional: Determine the rectilinear trajectories.

f) Sketch the phase portrait.

e) Determine the equations of the vertical and horizontal tangents.

f) Optional: Determine the rectilinear trajectories.

g) Sketch the phase portrait.

Reaction system

The following reaction system is given:

$$A + X \xrightarrow{k_1} 2X$$

$$X + Y \xrightarrow{k_2} 2Y$$

$$2Y \xrightarrow{k_3} B$$

The concentration of substance A shall be constant. The reactions take place in an ideally mixed enclosed reactor with constant volume.

a) Give the balance equations for the amount of substance (mol) of the reaction partners X and Y.

One obtains the following differential equation system by specifically choosing the reaction constants k_1, k_2, k_3:

$$\frac{dc_X}{dt} = c_X(1 - c_Y)$$

$$\frac{dc_Y}{dt} = c_Y(c_X - c_Y)$$

b) Determine the steady states of the system.

c) Linearize the differential equations around the steady state and transform them into state-space description (states, inputs, outputs in matrix/vector representation).

d) Calculate the Eigenvalues of the linearized system at the steady states and determine thereof the behavior of the system in the vicinity of the steady states.

Notes

4. Solutions

Do not betray yourself!

Exercises
■ Steady states
Calculate the steady states for systems 3.1 to 3.8:

Task 3.1

$$\dot{x} = 2x - 1$$
$$0 = 2x - 1$$
$$x^* = \frac{1}{2}$$

Task 3.2

$$\dot{x} = x^2 + 2x - 1$$
$$0 = x^2 + 2x - 1$$

You need the p-q equation (learn it by heart!) and then you get:

$$x^*_{1,2} = -\frac{2}{2} \pm \sqrt{\left(\frac{2}{2}\right)^2 + 1} = -1 \pm \sqrt{2}$$

Task 3.3

$$\dot{x} = \sin(x)$$
$$0 = \sin(x)$$
$$x^* = n\pi \quad \text{for: } n \in \mathbb{Z} \qquad \mathbb{Z} = \{\ldots; -2; -1; 0; 1; 2; \ldots\}$$

Works for **integers** $\mathbb{Z}$.

Task 3.4

$$\ddot{x} + \dot{x} + x = 1$$
$$x^* = 1$$

Task 3.5

$$\ddot{x} + 2\dot{x} + x^2 - 5x + 6 = 0$$
$$x^2 - 5x + 6 = 0$$
$$(x - 2)(x - 3) = 0$$
$$x^*_1 = 3$$
$$x^*_2 = 2$$

Alternatively, with p-q equation:

$$x^*_{1,2} = \frac{5}{2} \pm \sqrt{\left(\frac{5}{2}\right)^2 - 6}$$
$$= \frac{5}{2} \pm \frac{1}{2}$$

Task 3.6

$$\frac{d}{dt}\begin{pmatrix} x_1 \\ x_2 \end{pmatrix} = \begin{pmatrix} 3 & -2 \\ 1 & 5 \end{pmatrix}\begin{pmatrix} x_1 \\ x_2 \end{pmatrix} + \begin{pmatrix} 1 \\ 0 \end{pmatrix} u$$

is:

$$3x_1 - 2x_2 + u = 0 \qquad \text{I}$$
$$x_1 + 5x_2 = 0 \qquad \text{II}$$

$$\text{II}: \quad \Rightarrow \quad x_1 = -5x_2$$

in I:

$$-15x_2 - 2x_2 + u = 0$$
$$-17x_2 = -u$$
$$x^*_2 = \frac{1}{17}u$$

Back in II:

$$x_1 + \frac{5}{17}u = 0$$
$$x_1 = -\frac{5}{17}u$$

Thus, we have:

$$\begin{bmatrix} x^*_1 \\ x^*_2 \end{bmatrix} = \begin{pmatrix} -5/17 \\ 1/17 \end{pmatrix} u$$

or the steady state $(-5/12u, -1/17u)$.

Task 3.7

$$\dot{x} = (1 - x - y)x \qquad \text{I}$$
$$\dot{y} = (1 - x - y)y \qquad \text{II}$$

$$\text{I}: \quad \Rightarrow \quad 0 = (1 - x - y)x$$

<u>Case 1:</u>

$$\Rightarrow \boxed{x^*_1 = 0}$$

in **II**:

$$(1 - y)y = 0 \Rightarrow \boxed{Y^*_1 = 0} \text{ and } \boxed{Y^*_2 = 1}$$

The steady states are $(0; 0)$ & $(0; 1)$.

<u>Case 2:</u>

$$(1 - x - y) = 0$$
$$x = 1 - y$$
$$y = 1 - x$$

in **II** :

$$(1-1+y-y)y = 0$$

$$0 \cdot y = 0 \;\rightarrow\; \text{always true}$$

$$y^* = c \;\text{any desired value}$$

With this we have the steady states $(1-y;c)$ & $(c;1-x)$ which includes the steady state $(0;1)$.

Task 3.8

$$\dot{u} = a - (b+1)u + u^2 v$$

$$\dot{v} = bu - u^2 v$$

becomes:

$$\dot{u} = vu^2 - (b+1)u + a = 0 \qquad \text{I}$$

$$\dot{v} = u(vu - b) = 0 \qquad \text{II}$$

Case 1:

$$\text{II} \quad \text{if} \quad u = 0$$

$$\text{in I} \quad \Rightarrow \quad \boxed{a = 0} \qquad \text{\textreferencemark if } a \neq 0$$

Case 2:

$$\text{II} \quad \text{if} \quad u \neq 0$$

$$\Rightarrow \quad v = \frac{b}{u}$$

$$\Rightarrow bu - u - bu + a = 0$$

with the two solutions:

$$u^* = a$$

$$v^* = \frac{b}{a}$$

■ **Stability**
Test the following systems for stability of their steady state(s). Draw, therefore, $\dot{x}$ over x. See Figure 42.
Case 1: Stable

$$\text{if} \quad x > x^*$$

$$\text{then} \quad \dot{x} < 0$$

or

$$\text{if} \quad x < x^*$$

$$\text{than} \quad \dot{x} > 0$$

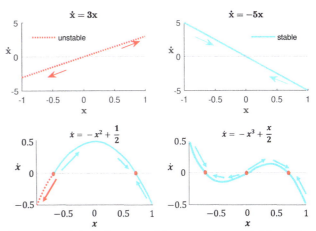

Figure 42. Plot of change of $\dot{x}$ over state x. Stable: if the state x is positive, the change $\dot{x}$ makes it more negative. Unstable: if the state x is positive, the change $\dot{x}$ makes it even more positive.

Case 2: Unstable

$$\text{if} \quad x > x^*$$

$$\text{then} \quad \dot{x} > 0$$

or

$$\text{if} \quad x < x^*$$

$$\text{then} \quad \dot{x} < 0$$

■ **Eigenvalues and stability of the steady state**
Task 3.13

$$\frac{d}{dt}\begin{pmatrix} x_1 \\ x_2 \end{pmatrix} = \begin{pmatrix} -1 & 4 \\ 0 & -3 \end{pmatrix}\begin{pmatrix} x_1 \\ x_2 \end{pmatrix}$$

$s1 = -1; s2 = -3$: stable

Task 3.14

$$\frac{d}{dt}\begin{pmatrix} x_1 \\ x_2 \end{pmatrix} = \begin{pmatrix} 3 & 8 \\ 1 & 1 \end{pmatrix}\begin{pmatrix} x_1 \\ x_2 \end{pmatrix}$$

$s1 = -1; s2 = 5$: unstable

Task 3.15

$$\frac{d}{dt}\begin{pmatrix} x_1 \\ x_2 \end{pmatrix} = \begin{pmatrix} -1 & 3 \\ -1 & 1 \end{pmatrix}\begin{pmatrix} x_1 \\ x_2 \end{pmatrix}$$

$s1 = i\sqrt{2}; s2 = -i\sqrt{2}$: metastable oscillations

■ **Population growth**
The growth of a specific population is described with the following model:

$$\frac{dX}{dt} = (2-X)(X-1)(X-k) \qquad (\text{with } k > 0 \text{ and } X(0) = X_0)$$

Task a) Derive the steady states of the system.

$$(2 - X^*)(X^* - 1)(X^* - k) = 0$$

$$\Rightarrow \boxed{X_1^* = 1}, \boxed{X_2^* = 2}, \boxed{X_3^* = k}$$

Task b) Linearize the differential equation around the steady states.

$$\dot{X} = (2X - 2 - X^2 + X)(X - k)$$
$$= (3X - 2 - X^2)(X - k)$$
$$= (3X^2 - 3kX - 2X + 2k - X^3 + kX^2)$$
$$= -X^3 + X^2(3 + k) + X(-3k - 2) + 2k$$
$$\Delta\dot{X} = \left[-3(X^*)^2 + 2X^*(3 + k) - 3k - 2\right] \cdot \Delta X$$

Task c) Calculate the Eigenvalues of the linearized system in the steady states and determine thereof the stability of the individual steady states. For which values of k do you get stable steady states?

Steady State 1 ($X^* = 1$):

$$\Rightarrow \quad s = -3 + 2(3 + k) - 3k - 2$$
$$= -3 + 6 + 2k - 3k - 2 = 1 - k$$

$\Rightarrow$ stable for $k > 1$.
Steady State 2 ($X^* = 2$):

$$\Rightarrow \quad s = -12 + 4(3 + k) - 3k - 2$$
$$= k - 2$$

$\Rightarrow$ stable for $k < 2$.
Steady State 3 ($X^* = k$):

$$\Rightarrow \quad s = -3k^2 + 2k(3 + k) - 3k - 2$$
$$= -3k^2 + 6k + 2k^2 - 3k - 2$$
$$= -k^2 + 3k - 2$$

$\Rightarrow$ stable

$$s = -k^2 + 3k - 2 < 0$$
$$k^2 - 3k + 2 > 0$$
$$(k - 1)(k - 2) > 0$$

Solution 1: $k - 1 > 0$ OR $k - 2 > 0$
$$\Rightarrow k > 2$$
Solution 2: $k - 1 < 0$ OR $k - 2 < 0$
$$\Rightarrow k < 1$$

Task d) and Task e) Mark within a X-over-t diagram the domains where $\dfrac{dX}{dt}$ is positive, then respectively negative. One parameter is fixed $k = 5$. Draw $X(t)$ in a diagram using the results of parts a)-d) for different initial conditions of X. One parameter is fixed $k = 5$.

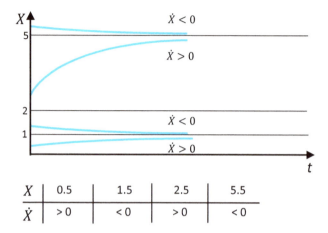

X	0.5	1.5	2.5	5.5
$\dot{X}$	> 0	< 0	> 0	< 0

◼ 1st-order system
The growth of a specific population is described with the following model:

$$\frac{dN}{dt} = rN(2 - N)(N - 1) \qquad \text{(with } r > 0 \text{ and } N(0) = N_0\text{)}$$

Task a) Derive the steady states of the system.

$$\frac{dN}{dt} = rN(2 - N)(N - 1) \overset{!}{=} 0 \Rightarrow \boxed{N_1^* = 0}; \boxed{N_2^* = 1}; \boxed{N_3^* = 2}$$

Task b) Linearize the differential equation around the steady states.
The equation can also be written in the following form:

$$\dot{N} = rN(2N - 2 - N^2 + N)$$
$$= rN(-N^2 + 3N - 2)$$
$$= -rN^3 + 3rN^2 - 2rN.$$

With only small deflections n, one can linearize ODEs around the steady states:

$$N = N_i^* + n \qquad \text{for } i = 1, 2, 3$$

According to Taylor:

$$\frac{d}{dt}(N^* + n) = \underbrace{f(N) \mid_{N_i^*}}_{=0} + \frac{\partial f(N)}{\partial N} \mid_{N_i^*} \cdot n + \dots$$

After the implementation of the non-linear equation into the Taylor equation, one obtains:

$$\frac{d}{dt}(N^* + n) = \underbrace{-r(N^*)^3 + 3r(N^*)^2 - 2rN^*}_{0}$$
$$+ ((-3(N^*)^2 + 6N^* - 2)r) \cdot n$$

$$\boxed{\frac{d}{dt}n = (-3(N^*)^2 + 6N^* - 2)rn}$$

Steady State 1: $N_1^* = 0 \rightarrow \frac{d}{dt}n = -2rn$

Steady State 2: $N_2^* = 1 \rightarrow \frac{d}{dt}n = rn$

Steady State 3: $N_3^* = 2 \rightarrow \frac{d}{dt}n = -2rn$

Task c) Calculate the Eigenvalues of the linearized system in the steady states and test for stability of the individual steady states.

Steady State 1: The Eigenvalue is s=-2r and is negative because $r > 0$. The Steady State 1 is thus asymptotically stable.

Steady State 2: Because the Eigenvalue s=r is always positive, the Steady State 2 is unstable.

Steady State 3: Similar to Steady State 1 and thus asymptotically stable.

Task d) Mark within a $t - N-$diagram the domains where $\frac{dN}{dt}$ is positive, then respectively negative.

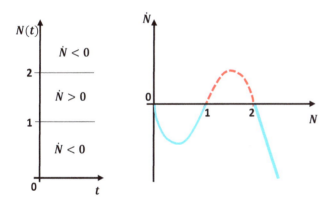

In the second plot ($\dot{N}$ over N) one can see very well where the derivative becomes negative or positive.

Task e) Draw $N(t)$ in a diagram using the results of parts a)-d) for different initial conditions of N. Calculate also the position of the turning points.
For the turning points, we need the second derivative:

$$\frac{d^2}{dt^2}N = \frac{d}{dN}f(N) \cdot \frac{d}{dt}N$$
$$= r[(2-N)(N-1) - N(N-1) + N(2-N)]\dot{N} = 0$$

This solution of the equation gives us only trivial solutions $\dot{N} = 0$. This does not mean that the steady states are inflection points at the same time, because the third

derivative is zero for equilibrium points. To sketch the curves we are only interested in the solutions:

$$r[(2-N)(N-1) - N(N-1) + N(2-N)] \overset{!}{=} 0$$
$$-3N^2 + 6N - 2 = 0$$
$$N^2 - 2rN + {}^2\!/\!{}_3 r = 0$$
$$N^2 - 2N + {}^2\!/\!{}_3 = 0$$

With:

$$\Rightarrow N_{1,2} = \frac{2 \pm \sqrt{4 - 8/3}}{2} = 1 \pm \frac{\sqrt{3}}{3}$$
$$\Rightarrow n_1 = 1.58, \quad n_2 = 0.42$$

we obtain the solution:

$$\boxed{N = 1 \pm \frac{\sqrt{3}}{3}}$$

Now one can sketch some solutions for the time-course behavior with $N(t)$ over t.

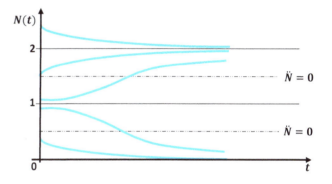

■ **Characteristic equation of a 2nd-order system**
The following characteristic equation of a 2nd-order system is given:

$$(s - a + T)(s - 1 + aT) = 0 \quad (a, T \text{ are real}) \qquad \text{I}$$

Task a) What is the corresponding homogeneous differential equation? Characteristic equation with formal ansatz:

$$y(t) = e^{st}$$
$$\dot{y}(t) = se^{st} = sy(t)$$

$$s^2 + (T - a + aT - 1)s + (T - a)(aT - 1) = 0$$
$$s^2 + (T - 1)(1 + a)s + (T - a)(aT - 1) = 0$$

Homogeneous ODE:

$$\ddot{x} + (T - 1)(1 + a)\dot{x} + (T - a)(aT - 1)x = 0$$

Task b) For which values of the parameters a and T do we have *one single* Eigenvalue on the stability bound?

From Equation I we get:

$$s_1 = a - T$$
$$s_2 = 1 - aT$$

An Eigenvalue lays on the stability margin, if:

1.) $\boxed{a = T} \to s_1 = 0, \quad s_2 = 1 - T^2 \neq 0 \quad \Rightarrow T \neq \pm 1$

2.) $\boxed{a = \dfrac{1}{T}} \to s_2 = 0, \quad s_1 = \dfrac{1}{T} \neq 0 \quad \Rightarrow T \neq \pm 1$

Task c) For which values of the parameters a and T is the system stable, metastable, or unstable?

1.) Asymptotic stability exists for $s_1 < 0$ and $s_2 < 0$.

$$a - T < 0 \quad \text{and} \quad 1 - aT < 0 \to aT > 1$$

$\boxed{a < T}$ Case 1: $\boxed{a > \dfrac{1}{T} \text{ AND } T > 0}$

                  Case 2: $\boxed{a < \dfrac{1}{T} \text{ AND } T < 0}$

2.) Metastability exists if one Eigenvalue lies on the stability margin and the rest is negative.

α) $a = T$ and $T^2 > 1$ $\to \boxed{T < -1 \text{ OR } T > 1}$

β) $a = \dfrac{1}{T}$ and $\dfrac{1}{T} - T < 0$ $\to aT > 1$

               Case 1: $\boxed{T > 0 \text{ AND } T > 1}$

               Case 2: $\boxed{T < 0 \text{ AND } T > -1}$

3.) In the remaining cases we have no stability.

Task d) Draw the results of c) in a stability diagram with abscissa T and ordinate a.

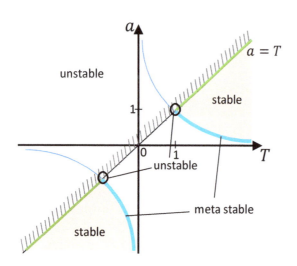

■ Biomass growth in a bioreactor

Task a) Set up balance equations for biomass (X) and substrate (S) concentrations.

$$\dot{X} = \mu X \qquad\qquad\qquad \text{I}$$
$$\dot{S} = -\frac{1}{Y_{XS}} \dot{X}$$
$$\dot{S} = -\frac{1}{Y_{XS}} \mu X \qquad\qquad \text{II}$$

Task b) Derive $S(t)$ as a function of $X(t)$, X_0, and S_0.

$$\text{Equation I} \Rightarrow \int_{X_0}^{X} \frac{dX}{X} = \int_0^t \mu \, dt$$
$$[\ln(X)]_{X_0}^{X} = [\mu \cdot t]_0^t$$
$$\ln(X) - \ln(X_0) = [\mu \cdot t]_0^t$$
$$\ln(X/X_0) = \mu t$$
$$\boxed{X = X_0 \cdot e^{\mu t}}$$
$$\text{Equation II} \Rightarrow \frac{dS}{dt} = -\frac{1}{Y_{XS}} \cdot \mu X_0 \cdot e^{\mu t}$$
$$dS = -\frac{1}{Y_{XS}} \cdot \mu X_0 \cdot e^{\mu t} \cdot dt$$
$$\int_{S_0}^{S} dS = -\frac{\mu X_0}{Y_{XS}} \cdot \int_0^t e^{\mu t} \cdot dt$$
$$[S]_{S_0}^{S} = -\frac{\mu X_0}{Y_{XS}} \cdot \frac{1}{\mu} \left[e^{\mu t} \right]_0^t$$
$$S - S_0 = -\frac{X_0}{Y_{XS}} \left(e^{\mu t} - 1 \right)$$
$$\boxed{S = S_0 - \frac{1}{Y_{XS}} [X(t) - X_0]}$$

or simply:

$$\frac{dS}{dt} = -\frac{1}{Y_{XS}} \cdot \mu X$$
$$= -\frac{1}{Y_{XS}} \cdot \frac{dX}{dt}$$
$$dS = -\frac{1}{Y_{XS}} \cdot dX$$
$$\int_{S_0}^{S(t)} = -\frac{1}{Y_{XS}} \cdot \int_{X_0}^{X(t)} dX$$
$$[S]_{S_0}^{S(t)} = -\frac{1}{Y_{XS}} \cdot [X]_{X_0}^{X(t)}$$
$$S(t) - S_0 = -\frac{1}{Y_{XS}} \cdot [X(t) - X_0]$$
$$\Rightarrow S(t) = S_0 - \frac{1}{Y_{XS}} \cdot [X(t) - X_0]$$

Task c) Derive a differential equation for X which does not depend on the substrate concentration anymore [$\dot{X} =$

$f(X)$] using the assumption $\mu = cS$.

$$\frac{dX}{dt} = \mu X = cSX$$

$$= c[S_0 - \frac{1}{Y_{XS}}[X(t) - X_0]]X$$

Task d) Repeat this while assuming a logistic growth rate $\mu = a - bX$ with $(a, b > 0)$.

$$\frac{dX}{dt} = \mu X = X(a - bX) = aX - bX^2 \qquad (4.1)$$

$$\frac{dS}{dt} = -\frac{1}{Y_{XS}} \cdot X(a - bX) \qquad (4.2)$$

Task e) Compare the differential equations obtained for X in c) and d). What values would you have to choose for a and b in order to obtain identical differential equations?

$$\frac{dX}{dt} = c\underbrace{\left[S_0 - \frac{X}{Y_{XS}} + \frac{X_0}{Y_{XS}}\right]X}_{\text{Task c)}} = \underbrace{aX - bX^2}_{\text{Task d)}} \overset{!}{=} 0$$

$$0 = X\left[cS_0 + \frac{cX_0}{Y_{XS}} - \frac{cX}{Y_{XS}}\right] = X[a - bX]$$

A comparison of coefficients delivers:

$$\Rightarrow a = c\left(S_0 + \frac{X_0}{Y_{XS}}\right)$$

$$\Rightarrow b = \frac{c}{Y_{XS}}$$

Task f) Calculate the steady states (X^*, S^*) in task d).

$$0 = (a - bX)X \qquad (4.3)$$

$$0 = -\frac{a - bX}{Y_{XS}}X \qquad (4.4)$$

Equation 4.3 delivers two solutions:

$$X_1^* = 0 \qquad X_2^* = \frac{a}{b}$$

Inserting into Equation 4.4

$$0 = -\frac{a - b0}{Y_{XS}}0 \quad \text{and} \quad 0 = -\frac{a - b\frac{a}{b}}{Y_{XS}}\frac{a}{b} = -\frac{0}{Y_{XS}}\frac{a}{b}$$

shows that the substrate does not matter and that the steady-state values for X are valid for any substrate concentration. The solution does not say anything about the dynamic until the steady state. One could say that either the reactor content is dead at the end or the growth and death rate balances the amount of individuals. Have in mind that the logistic growth equation does not depend on the "substrate" and it is the responsibility of the modeler to set up realistic but still simple equations.

Task g) Show that the following ansatz is a solution for $X(t)$:

$$X(t) = \frac{X^*}{1 + c_1 e^{-c_2 t}}$$

What values would you have to choose for c_1 and c_2?

The equation:

$$\frac{dX}{dt} = (a - bX)X$$

has X itself and its derivative. We do not have the derivative of the ansatz yet. Let's do it! Therefore, we need the quotient rule:

$$\left(\frac{f}{g}\right)' = \frac{f'g - fg'}{g^2}$$

which is applied on our ansatz:

$$\frac{dX(t)}{dt} = \frac{\left(0 \cdot \frac{X^*}{1 + c_1 e^{-c_2 t}}\right) - (X^*(0 + c_1 c_2 e^{-c_2 t}))}{(1 + c_1 e^{-c_2 t})^2}$$

$$= \frac{X^* c_1 c_2 e^{-c_2 t}}{(1 + c_1 e^{-c_2 t})^2}$$

Now we have an equation for X and $\dot{X}$. We can insert the ansatz into Equation 4.1.

$$\frac{dX}{dt} = (a - bX)X$$

$$\frac{X^* c_1 c_2 e^{-c_2 t}}{(1 + c_1 e^{-c_2 t})^2} = \left(a - b\frac{X^*}{1 + c_1 e^{-c_2 t}}\right)\frac{X^*}{1 + c_1 e^{-c_2 t}}$$

and multiply with $(1 + c_1 e^{-c_2 t})^2$:

$$X^* c_1 c_2 e^{-c_2 t} = \left(a - b\frac{X^*}{1 + c_1 e^{-c_2 t}}\right)\frac{X^*(1 + c_1 e^{-c_2 t})^2}{1 + c_1 e^{-c_2 t}}$$

$$X^* c_1 c_2 e^{-c_2 t} = (a(1 + c_1 e^{-c_2 t}) - b)X^*$$

$$0 = (a(1 + c_1 e^{-c_2 t}) - bX^* - c_1 c_2 e^{-c_2 t})X^*$$

$$= (a - bX^* + (a - c_2)c_1 e^{-c_2 t})X^*$$

Check the steady states

Let's test the first steady state $\boxed{X^* = 0}$:

$$0 = (a - b \cdot 0 + (a - c_2)c_1 e^{-c_2 t}) \cdot 0$$

Well, this equation is always true. But we are rather interested in the time response.

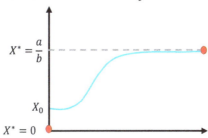

We could try to get more out of the ansatz by checking what happens at time point zero $t = 0$.

$$X(t = 0) = X_0 = \frac{X^*}{1 + c_1 e^{-c_2 0}} = \frac{X^*}{1 + c_1} \qquad (4.5)$$

which tells us that with the steady state $X^* = 0$:

$$X_0 = \frac{0}{1 + c_1}$$

the equation is only valid if $X_0 = 0$: So it means that if we do not have any micro-organisms in the tank at the beginning, we will also have no micro-organisms at the end and stay in the steady state $X* = 0$. Not really interesting, is it? Remark: this steady state is unstable. Let's do something useful by checking the second steady state.

We introduce $\boxed{X^* = \frac{a}{b}}$:

$$0 = \left(a - b \cdot \frac{a}{b} + (a - c_2)c_1 e^{-c_2 t} \right) \cdot \frac{a}{b}$$

with which we know that:

$$c_2 = a$$

We insert this into Equation 4.5:

$$X_0 = \frac{\frac{a}{b}}{1 + c_1}$$

$$1 + c_1 = \frac{\frac{a}{b}}{X_0}$$

$$c_1 = \frac{a}{bX_0} - 1$$

Inserting this into our ansatz

$$X(t) = \frac{X^*}{1 + c_1 e^{-c_2 t}}$$

we obtain a final solution:

$$X(t) = \frac{a/b}{1 + \left(\frac{a}{bX_0} - 1 \right) e^{-at}}$$

Isn't this beautiful?

■ **Prey-Predator Model**

We have 2 versions of the Prey-Predator Model of Lotka and Volterra. Prey X grows independently of Predator Y and is depleted by interacting with Y. Thereby Predator Y grows and, in turn, is depleted by a natural death rate. The most simple differential equation model reflecting these relations is:

$$\dot{X} = X(1 - Y) \qquad \text{I}$$
$$\dot{Y} = Y(X - 1) \qquad \text{II}$$

Task a)   Assign the facts described above to the individual model terms in the system.

$$\text{Prey number change:} \quad \dot{X} = \underset{\text{birth}}{X} \quad \underset{\text{Predator eats prey}}{-YX}$$

$$\text{Predator number change:} \quad \dot{Y} = \underset{\text{growth on prey}}{YX} \quad \underset{\text{death}}{-Y}$$

Task b)   Calculate the steady states of the system.

$$\dot{X} = X^*(1 - Y^*) \overset{!}{=} 0 \qquad \text{I}$$
$$\dot{Y} = Y^*(X^* - 1) \overset{!}{=} 0 \qquad \text{II}$$

<u>Case differentiation:</u>

Equation I $\Rightarrow \boxed{X_1^* = 0}$

in Eq. II: $Y^*(0 - 1) = 0 \Rightarrow \boxed{Y_1^* = 0}$

Equation I $\Rightarrow \boxed{Y_2^* = 1}$

in Eq. II: $1 \cdot (X^* - 1) = 0 \Rightarrow \boxed{X^* = 1}$

Task c)   How can the behavior in vicinity of the steady states be approximated (without calculations)? Calcu-

Cases	$0 < Y < 1$	$Y > 1$
$0 < X < 1$	$\dot{X} > 0$ $\dot{Y} < 0$	$\dot{X} < 0$ $\dot{Y} < 0$
$X > 1$	$\dot{X} > 0$ $\dot{Y} > 0$	$\dot{X} < 0$ $\dot{Y} > 0$

lation close to the steady state at the zero point: (0.1 | 0.1):

$$\Rightarrow \dot{X} = 0.1(1 - 0.1) = 0.09$$
$$\dot{Y} = 0.1(0.1 - 1) = -0.09$$

Phase plot with oscillating system

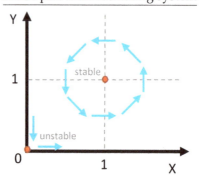

■ Prey-Predator Model 2

In the following, a more complex Lotka-Voltera model shall be investigated. The growth rate of the prey is included with a parameter $\mu > 0$ and the predator's growth rate is modeled with a yield coefficient $\alpha > 0$. Furthermore, an influx of predators into the system is assumed. The resulting model description is:

$$\dot{X} = X(\mu - Y) \qquad (4.6)$$
$$\dot{Y} = Y(\alpha X - 1) + 1 \qquad (4.7)$$

Task a) Assign the facts described above to the individual model terms in this system.

$$\text{Prey number change:} \quad \dot{X} = \underset{\text{birth}}{\mu X} \ \underset{\text{predation}}{-YX}$$

$$\text{Predator number change:} \quad \dot{Y} = \underset{\text{growth on prey}}{\alpha YX} \ \underset{\text{death}}{-Y} \ \underset{\text{immigration}}{+1}$$

Task b) Calculate the steady states of system as a function of parameters α and μ.

$$\dot{X} = X^*(\mu - Y^*) \overset{!}{=} 0 \qquad\qquad \text{I}$$
$$\dot{Y} = Y^*(\alpha X^* - 1) + 1 \overset{!}{=} 0 \qquad \text{II}$$

Case differentiation:

Equation I $\Rightarrow \boxed{X_1^* = 0}$

in Eq. II: $Y^*(0-1)+1 = 0 \Rightarrow \boxed{Y_1^* = 1}$

Equation I $\Rightarrow (\mu - Y^*) = 0 \Rightarrow \boxed{Y_2^* = \mu}$

in Eq. II: $\mu \cdot (\alpha X^* - 1) + 1 = 0$

$$\mu \alpha X^* - \mu + 1 = 0$$

$$\Rightarrow \boxed{X^* = \frac{\mu - 1}{\alpha \mu}}$$

Task c) Which condition has to be fulfilled for parameter μ to obtain only biologically reasonable steady states? We assume it is reasonable to always have enough of each species in in the system $X^*, Y^* \overset{!}{\geq} 0$ then:

$$Y^* \geq 0 \ \Rightarrow \qquad\qquad \mu \geq 0$$
$$X^* \geq 0 \ \Rightarrow \qquad \frac{\mu - 1}{\alpha \cdot \mu} \geq 0$$
$$\mu - 1 \geq 0$$
$$\boxed{\mu \geq 1}$$

Task d) Draw the steady states for $\mu = 2$ and $\alpha = [1/5;$ $1/4; 1/3; 1/2; 1]$ in a diagram (x-axis: X, y-axis: Y).

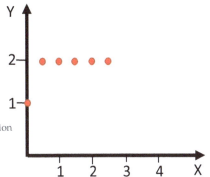

$\alpha =$	$X^* = \frac{\mu-1}{\alpha \cdot \mu} = \frac{1}{2\alpha}$	$Y^* = \mu$
$\frac{1}{5}$	$5/2$	2
$\frac{1}{4}$	2	2
$\frac{1}{3}$	$3/2$	2
$\frac{1}{2}$	1	2
1	$1/2$	2

Plot with parameter-dependent shift of the steady state:

■ Prey-Predator Model 3 (fish)

Small and big fishes are living in a lake. The small fishes (X) depend on plankton which is available in excess. Big fishes (Y) feed on the small ones. If only small fishes would be present, it would grow exponentially, *i.e.* their number grows with a constant specific growth rate α $(\alpha > 0)$. In the absence of small fishes, the big fishes would die out with a decay rate β $(\beta > 0)$. The following simple equations describe these facts using the parameters $\alpha, \beta, \gamma, \delta > 0$:

$$\frac{d}{dt}X = (\alpha - \gamma Y)X \overset{!}{=} 0 \qquad \text{I}$$
$$\frac{d}{dt}Y = -(\beta - \delta X)Y \overset{!}{=} 0 \qquad \text{II}$$

Task a) Determine the steady states of this simple ecological model.

Equation I $\Rightarrow \boxed{X_1^* = 0}$

in Eq. II: $-(\beta - \delta 0)Y = 0 \Rightarrow \boxed{Y_1^* = 0}$

Equation I $\Rightarrow (\alpha - \gamma Y) = 0 \Rightarrow \boxed{Y_2^* = \frac{\alpha}{\gamma}}$

in Eq. II: $-(\beta - \delta X)\frac{\alpha}{\gamma} = 0$

$$\delta X \frac{\alpha}{\gamma} = \beta \frac{\alpha}{\gamma}$$

$$\Rightarrow \boxed{X^* = \frac{\beta}{\delta}}$$

Task b) Linearize the differential equations around the steady states and transform the linearized equations into

state-space description (states, inputs, outputs in matrix/vector representation).
We assume:

$$X = X^* + x$$
$$Y = Y^* + y$$

The right-hand side of the non-linear equation system will be linearized around the steady states. If one denotes the right-hand side of the first ordinary differential equation (ODE) as function of X and Y, one can use the Taylor approximation:

$$\frac{d}{dt}X = \frac{d}{dt}(X^* + x)$$

$$= g(X^*, Y^*) + \frac{\partial g(X,Y)}{\partial X}\Big|_{X_S} \cdot x + \frac{\partial g(X,Y)}{\partial Y}\Big|_{Y_S} \cdot y + \dots$$

Applied on our system, we obtain:

$$\frac{d}{dt}x = (\alpha - \gamma Y^*) \cdot x - \gamma X^* \cdot y$$

$$\frac{d}{dt}y = \delta Y^* \cdot x - (\beta - \delta X^*) \cdot y$$

The related state-space representation is:

$$\begin{pmatrix} \dot{x} \\ \dot{y} \end{pmatrix} = \begin{pmatrix} \alpha - \gamma Y^* & -\gamma X^* \\ \delta Y^* & -\beta + \delta X^* \end{pmatrix} \begin{pmatrix} x \\ y \end{pmatrix}$$

Steady State 1: $X_1^* = 0, Y_1^* = 0$. The steady-state values are inserted into the state-space representation:

$$\begin{pmatrix} \dot{x} \\ \dot{y} \end{pmatrix} = \begin{pmatrix} \alpha & 0 \\ 0 & -\beta \end{pmatrix} \begin{pmatrix} x \\ y \end{pmatrix} \tag{4.8}$$

Steady State 1: $X_2^* = \frac{\beta}{\delta}, Y_2^* = \frac{\alpha}{\gamma}$

$$\begin{pmatrix} \dot{x} \\ \dot{y} \end{pmatrix} = \begin{pmatrix} 0 & -\gamma\frac{\beta}{\delta} \\ -\delta\frac{\alpha}{\gamma} & 0 \end{pmatrix} \begin{pmatrix} x \\ y \end{pmatrix} \tag{4.9}$$

Task c) Determine for every steady state the Eigenvalues of the linearized system and deduce from that the stability of the system in the vicinity of the respective steady state.
Steady State 1:
Eigenvalues directly from the related state-space representation 4.8: $s_1 = \alpha, s_2 = -\beta \rightarrow$ steady state is unstable
Steady State 2:
We use the state-space representation 4.9 to get the characteristic equation:

$$|sI - A| = \begin{vmatrix} 0 & -\gamma\frac{\beta}{\delta} \\ -\delta\frac{\alpha}{\gamma} & 0 \end{vmatrix} = s^2 + \alpha\beta = 0$$

Thus, we obtain the complex Eigenvalues:

$$s_{1,2} = \pm\sqrt{-\alpha\beta} = \pm i\sqrt{\alpha\beta}$$

Thus, we have a metastable steady state with conserved oscillations, also known as a limit cycle.

■ **Prey-Predator Model 4 (fish)**

$$g_1(x,y) = \dot{x} = (1-y)x - \frac{1}{2}x^2 = x - xy - \frac{1}{2}x^2 \overset{!}{=} 0 \qquad \text{I}$$

$$g_2(x,y) = \dot{y} = (x-1)y \qquad = -y + xy \qquad \overset{!}{=} 0 \qquad \text{II}$$

Task a) Determine the steady state.
From II:

$$(x-1)y = 0 \rightarrow \boxed{y_1^* = 0}$$
$$\boxed{x_2^* = 1}$$

$y_1^* = 0$ in I:

$$x(1 - \frac{1}{2}x) = 0 \rightarrow \boxed{x_1^* = 0}$$
$$\boxed{x_3^* = 2} \rightarrow \boxed{y_3^* = 0}$$

$x_2^* = 1$ in I:

$$1 - y - \frac{1}{2} = 0 \rightarrow \boxed{y_2^* = \frac{1}{2}}$$

Task b) Linearize around the steady state with the equation

$$f(x,y) = g(x^*, y^*) + g_x'(x^*, y^*)\Delta x + g_y'(x^*, y^*)\Delta y$$

using $x = x^* + \Delta x$ with $\Delta x = (x - x^*)$.

$$\dot{x} = f_1(x^*, y^*) = (x^* - x^*y^* - \frac{1}{2}(x^*)^2) + (1 - y^* - x^*)\Delta x + (-x^*)\Delta y$$

$$\dot{y} = f_2(x^*, y^*) = (-y^* + x^*y^*) + (y^*)\Delta x + (x^* - 1)\Delta y$$

which is in the general state-space representation:

$$\begin{pmatrix} \dot{x} \\ \dot{y} \end{pmatrix} = \underbrace{\begin{pmatrix} 1 - y^* - x^* & -x^* \\ y^* & x^* - 1 \end{pmatrix}}_{J = \frac{\partial g(x)}{\partial x}} \begin{pmatrix} \Delta x \\ \Delta y \end{pmatrix} + \begin{pmatrix} x^* - x^*y^* - \frac{1}{2}(x^*)^2 \\ -y^* + x^*y^* \end{pmatrix}$$

and for Steady State 1 $(x = 0 | y = 0) \rightarrow (0|0)$:

$$\begin{pmatrix} \dot{x} \\ \dot{y} \end{pmatrix} = \begin{pmatrix} 1 & 0 \\ 0 & -1 \end{pmatrix} \begin{pmatrix} \Delta x \\ \Delta y \end{pmatrix},$$

Steady State 2 $(1|\frac{1}{2})$:

$$\begin{pmatrix} \dot{x} \\ \dot{y} \end{pmatrix} = \begin{pmatrix} -\frac{1}{2} & -1 \\ \frac{1}{2} & 0 \end{pmatrix} \begin{pmatrix} \Delta x \\ \Delta y \end{pmatrix},$$

Steady State 3 $(2|0)$:

$$\begin{pmatrix} \dot{x} \\ \dot{y} \end{pmatrix} = \begin{pmatrix} -1 & -2 \\ 0 & 1 \end{pmatrix} \begin{pmatrix} \Delta x \\ \Delta y \end{pmatrix},$$

Task c)   Determine the stability with the Eigenvalues:

$$|s\mathbf{I} - A| = \begin{vmatrix} s - a_{11} & a_{12} \\ a_{21} & s - a_{22} \end{vmatrix}$$

$$= (s - a_{11})(s - a_{22}) - a_{12}a_{21} \overset{!}{=} 0$$

For <u>Steady State 1</u> $(0|0)$:

$$(s - 1)(s + 1) \overset{!}{=} 0$$

This characteristic polynomial delivers $s_1 = 1$, $s_2 = -1$, which indicates an *unstable saddle point*.
Asymptote:
We start with the equation in the state plane:

$$\frac{\dot{y}}{\dot{x}} = \frac{dy}{dx} = \frac{-y}{x}$$

and introduce the linear equation $y = mx$ and obtain a equation

$$m = \frac{-mx}{x}$$

that only works for $m = 0$

<u>Steady State 2</u> $(1|\frac{1}{2})$ delivers the characteristic equation:

$$(s + \frac{1}{2}) \cdot s + \frac{1}{2} = s^2 + \frac{1}{2}s - \frac{1}{2} = 0$$

$$2s^2 + s + 1 = 0$$

$$s_{1/2} = \frac{-1 \pm \sqrt{1 - 8}}{4} = -\frac{1}{4} \pm i\frac{\sqrt{7}}{4}$$

which indicates a *stable focus*.
Asymptote:
We start with the equation in the state plane

$$\frac{\dot{y}}{\dot{x}} = \frac{dy}{dx} = \frac{\frac{1}{2}x}{-\frac{1}{2}x - y}$$

and introduce the linear equation $y = mx$ and obtain the equation

$$m = \frac{\frac{1}{2}x}{-\frac{1}{2}x - mx}$$

$$-\frac{1}{2}mx - m^2x = \frac{1}{2}x$$

$$0 = m^2 + \frac{1}{2}m + \frac{1}{2}$$

$$m_{1,2} = -\frac{1}{4} \pm \sqrt{\frac{1}{16} - \frac{1}{2}} = \frac{-1 \pm \sqrt{-7}}{4}$$

which delivers imaginary slope values.
<u>Steady State 3</u> $(2|0)$ result in the characteristic equation

$$(s + 1)(s - 1) - 0 \overset{!}{=} 0 \quad \Rightarrow \quad s_1 = 1; s_2 = -1$$

and consequently is an *unstable saddle.*
Asymptote:
We start with the equation in the state plane

$$\frac{\dot{y}}{\dot{x}} = \frac{dy}{dx} = \frac{y}{-x - 2y}$$

and introduce the linear equation $y = mx$ and obtain a equation

$$m = \frac{mx}{-x - 2mx}$$

$$-mx - 2m^2x = mx$$

$$-2mx - 2m^2x = 0$$

$$2mx(1 + m) = 0$$

which result in the slope values $m_1 = 0$ and $m_2 = -1$

Task d)   Tangents
To draw our phase plot, we calculate horizontal and vertical asymptotes with:

- horizontal tangents: $\dot{y} \overset{!}{=} 0$ in Equation II:
  $\rightarrow x = 1$, $y = 0$

- vertical tangents: $\dot{x} \overset{!}{=} 0$ in Equation I:
  $\rightarrow x = 0$, $y = -\frac{1}{2}x + 1$

Task e)   Rectilinear tangents
Ansatz[10] [11]: $y = mx + b$

$$\frac{dy}{dx} = m = \frac{-y + xy}{x - xy - \frac{1}{2}x^2} = \frac{-mx - b + mx^2 + bx}{x - mx^2 - bx - \frac{1}{2}x^2} \quad (4.10)$$

---

[10] **Ansatz:** $x = my + b$

$$\frac{dx}{dy} = m = \frac{x - xy - \frac{1}{2}x^2}{-y + xy}$$

$$= \frac{my + b - my^2 - by - \frac{1}{2}[m^2y^2 + 2mby + b^2]}{-y + my^2 + by}$$

$$\Rightarrow -my + m^2y^2 + mby - my - b + my^2 + by + \frac{1}{2}m^2y^2 + mby + \frac{1}{2}b^2 = 0$$

$$\Rightarrow \underbrace{(m^2 + m + \frac{1}{2}m^2)}_{0!} y^2 + \underbrace{(b)}_{0!} y + \underbrace{(\frac{1}{2}b^2 - b)}_{0!} = 0$$

Coefficients:

$$b_1 = 0 \rightarrow m_1 = 0 \rightarrow \boxed{x = 0}$$

$$b_2 = 2 \rightarrow m_2 = 1 \rightarrow \frac{3}{2} + 1 = \frac{5}{2} \neq 0 \quad \text{⸘}$$

[11] **Ansatz:** $x = b$

$$x = b$$
$$\dot{x} = 0$$

$$\Rightarrow \dot{x} = (1 - y)x - \frac{1}{2}x^2 = x(1 - y - \frac{x}{2}) \overset{!}{=} 0$$

$$\Rightarrow b(1 - y - \frac{b}{2}) = 0 \quad \Rightarrow b = 0 \quad \text{(for all y)} \quad \Rightarrow \boxed{x = 0}.$$

$$\Rightarrow mx - m^2x^2 - bmx - \frac{1}{2}mx^2 = -mx - b + mx^2 + bx$$

$$\Rightarrow -m^2x^2 - \frac{1}{2}mx^2 - mx^2 + mx - bmx + mx - bx + b = 0$$

$$\Rightarrow \underbrace{(-\frac{3}{2}m - m^2)}_{0!}x^2 + \underbrace{(2m - bm - b)}_{0!}x + \underbrace{b}_{0!} = 0$$

Coefficients:

$$\Rightarrow \boxed{b = 0}$$

$$\Rightarrow 2m - 0 \cdot m - 0 = 0 \rightarrow 2m = 0 \rightarrow \boxed{m = 0}$$

$$\Rightarrow m(m + \frac{3}{2}) = 0 \rightarrow \boxed{m_1 = 0;\ m_2 = -\frac{3}{2}}$$

Inserted in Equation 4.10:

for $m_1 = 0 \Rightarrow \boxed{y = 0}$

for $m_1 = -\frac{3}{2}$

$$-\frac{3}{2} = \frac{\frac{3}{2}x - \frac{3}{2}x^2}{x + x^2}$$

$$-\frac{3}{2} = \frac{3}{2} \quad \text{↯}$$

Task f) Draw the phase plot.
The phase plot can be seen in Figure 43.

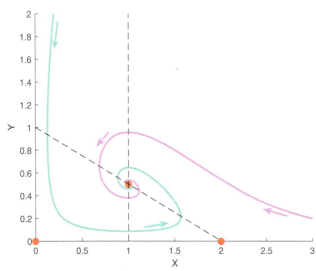

Figure 43. Phase plot: Red dots are steady states of the system. Dotted lines are the vertical and horizontal tangents. Initial values are: $(x = 0.2, y = 2)$ and $(x = 2, y = 0.2)$.

■ **Reaction system**
Task a) Balance equations for X and Y.
The balance equation can be written directly

$$\dot{A} = 0$$
$$\dot{X} = -k_1AX + 2k_1AX - k_2XY = X(k_1A - k_2Y)$$
$$\dot{Y} = -k_2XY + 2k_2XY - 2k_3Y^2 = Y(k_2X - 2k_3Y)$$
$$\dot{B} = +k_3Y^2$$

whereby one has to carefully pay attention to the coefficients. Or one uses the network approach learned in the previous section:

$$N = \begin{matrix} & v_1 & v_2 & v_3 & \\ & \begin{bmatrix} -1 \rightarrow 0 & 0 & 0 \\ -1 + 2 & -1 & 0 \\ 0 & -1 + 2 & -2 \\ 0 & 0 & 1 \end{bmatrix} & \begin{matrix} A \\ X \\ Y \\ B \end{matrix} \end{matrix}$$

which results in:

$$\begin{bmatrix} \dot{A} \\ \dot{X} \\ \dot{Y} \\ \dot{B} \end{bmatrix} = \dot{S} = Nv = \begin{bmatrix} 0 & 0 & 0 \\ 1 & -1 & 0 \\ 0 & 1 & -2 \\ 0 & 0 & 1 \end{bmatrix} \begin{bmatrix} k_1AX \\ k_2XY \\ k_3Y^2 \end{bmatrix}.$$

This way might be less error prone in praxis. In both cases we obtain:

$$\dot{X} = X(k_1A - k_2Y)$$
$$\dot{Y} = Y(k_2X - 2k_3Y)$$

where A is kept constant, and B does not influence the behavior of X or Y. Because A is constant, we could also combine it with k_1 to another constant $k_1A = C$. Also k_3 could be updated $2k_3 = k_3'$. In the following we do not want to drag the parameters on and only study the system:

$$\dot{X} = X(1 - Y)$$
$$\dot{Y} = Y(X - Y)$$

Task b) Determine the steady states of the system.

$$g_1(X, Y) = \dot{X} = X(1 - k_2Y) \qquad \overset{!}{=} 0 \qquad \text{I}$$
$$g_2(X, Y) = \dot{Y} = Y(X - Y) \qquad \overset{!}{=} 0 \qquad \text{II}$$

Steady State 1:

$$\boxed{x_1^* = 0} \Rightarrow \boxed{y_1^* = 0}$$

Steady State 2:

$$\boxed{y_2^* = 1}$$

in II

$$1(x - 1) \overset{!}{=} 0 \Rightarrow \boxed{x_2^* = 1}$$

Task c) Linearize the differential equations around the steady state and transform them into state-space descriptions (states, inputs, outputs in matrix/vector representation).
The general form is:

$$f(X,Y) = g(X^*,Y^*) + g_x'(X^*,Y^*)\Delta X + g_y'(X^*,Y^*)\Delta Y$$

sometimes also written as:

$$\Delta \dot{X} = f_1(X,Y) - g_1(X^*,Y^*)$$
$$= g_{1x}'(X^*,Y^*)\Delta X + g_{1y}'(X^*,Y^*)\Delta Y$$
$$\Delta \dot{Y} = f_2(X,Y) - g_2(X^*,Y^*)$$
$$= g_{2x}'(X^*,Y^*)\Delta X + g_{2y}'(X^*,Y^*)\Delta Y$$

using $X = X^* + \Delta X$ with $\Delta X = (X - X^*)$. In our example, we obtain:

$$\dot{X} = f_1(X^*,Y^*) = X^*(1-Y^*) + (1-Y^*)\Delta x + (-X^*)\Delta y$$
$$\dot{Y} = f_2(X^*,Y^*) = Y^*(X^*-Y^*) + (Y^*)\Delta x + (X^*-2Y^*)\Delta y$$

which is in the general state-space representation:

$$\begin{pmatrix} \dot{x} \\ \dot{y} \end{pmatrix} = \underbrace{\begin{pmatrix} 1-Y^* & -X^* \\ Y^* & X^*-2Y^* \end{pmatrix}}_{J = \frac{\partial \mathbf{g}(x)}{\partial x}} \begin{pmatrix} \Delta x \\ \Delta y \end{pmatrix} + \begin{pmatrix} X^*(1-Y^*) \\ Y^*(X^*-Y^*) \end{pmatrix}$$

Steady State 1 $(X^* = 0, Y^* = 0)$:

$$\begin{pmatrix} \dot{X} \\ \dot{Y} \end{pmatrix} = \begin{pmatrix} 1 & 0 \\ 0 & 0 \end{pmatrix} \begin{pmatrix} \Delta x \\ \Delta y \end{pmatrix}$$

Steady State 2 $(X^* = 1, Y^* = 1)$:

$$\begin{pmatrix} \dot{X} \\ \dot{Y} \end{pmatrix} = \begin{pmatrix} 0 & -1 \\ 1 & -1 \end{pmatrix} \begin{pmatrix} \Delta x \\ \Delta y \end{pmatrix}$$

Task d) Calculate the Eigenvalues of the linearized system at the steady states and determine thereof the behavior of the system in the vicinity of the steady states.

$$|s\mathbf{I} - A| = \begin{vmatrix} s - a_{11} & a_{12} \\ a_{21} & s - a_{22} \end{vmatrix}$$
$$= (s-a_{11})(s-a_{22}) - a_{12}a_{21} \overset{!}{=} 0$$

Steady State 1:

$$|s\mathbf{I} - A| = \begin{vmatrix} s-1 & 0 \\ 0 & s \end{vmatrix}$$
$$= (s-1)s \overset{!}{=} 0$$

with the Eigenvalues:

$$s_1 = 1$$
$$s_2 = 0$$

The positive Eigenvalue indicates an unstable steady state. Steady State 2:

$$|s\mathbf{I} - A| = \begin{vmatrix} s & 1 \\ -1 & s+1 \end{vmatrix}$$
$$= s(s+1) + 1 \overset{!}{=} 0$$
$$= s^2 + s + 1$$

The quadratic equation is solved:

$$s_{1,2} = \frac{1 \pm \sqrt{1-4}}{2} = -\frac{1}{2} \pm \frac{i}{2}\sqrt{3}$$

and results in 2 complex numbers with negative real part. Thus, we have a stable focus.

Task e) Determine the equations of the vertical and horizontal tangents.

Horizontal tangent:

$$\dot{Y} = Y(X-Y) \overset{!}{=} 0$$
$$\Rightarrow \boxed{Y = 0}; \quad \boxed{X = Y}$$

Vertical tangent:

$$\dot{X} = X(1-Y) \overset{!}{=} 0$$
$$\Rightarrow \boxed{X = 0}; \quad \boxed{Y = 1}$$

Task f) Optional: Determine the rectilinear trajectories. Ansatz[12][13]: $y = mx + b$

[12] Ansatz: $X = mY + b$

$$\frac{dX}{dY} = m = \frac{mY+b-(mY+b)Y}{(mY+b)Y - Y^2}$$
$$= \frac{mY+b-mY^2-bY}{mY^2+bY-Y^2}$$

$$\Rightarrow m^2Y^2 + bmY - mY^2 - mY - b + mY^2 + bY = 0$$
$$Y^2(m^2-m+m) + Y(bm-m+b) - b = 0$$

$$\Rightarrow \boxed{b=0}, \quad \boxed{m=0}, \quad \boxed{X=0}.$$

[13] Ansatz: $X = b$

$$\dot{X} = 0$$
$$\dot{X} = 0 = X(1-Y) = b(1-Y)$$
$$\Rightarrow b = 0, \quad \boxed{X=0}.$$

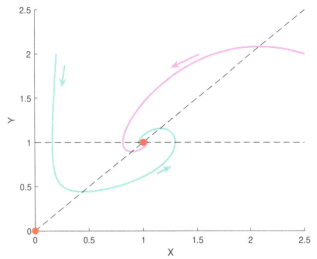

Figure 44. Phase plot: Red dots are steady states of the system. Dotted lines are the vertical and horizontal tangents. Initial values are: $(x = 0.2, y = 2)$ and $(x = 2.5, y = 2)$.

$$\frac{dY}{dX} = m = \frac{X(mX + b) - (mX + b)^2}{X - X(mX + b)}$$

$$\frac{mX^2 + bX - m^2X^2 - 2bmX - b^2}{X - mX^2 - bX}$$

$$\Rightarrow mX - m^2X^2 - bmX - mX^2 - bX + m^2X^2 + 2bmX + b^2 = 0$$

$$X^2(-m^2 - m + m^2) + X(m - bm - b + 2bm) + b^2 = 0$$

$$\Rightarrow \boxed{b = 0}, \quad \boxed{m = 0}, \quad \boxed{Y = 0}$$

Task g) Sketch the phase portrait.
We try to help our drawing with additional point tests:

Test at $(X = 2, Y = 1)$:

$$\dot{X} = 2(1 - 1) = 0$$
$$\dot{Y} = 1(2 - 1) = 1$$

Test at $(X = 1, Y = 2)$:

$$\dot{X} = 1(1 - 2) = -1$$
$$\dot{Y} = 2(1 - 2) = -2$$

The phase plot can be seen in Figure 44 and the corresponding MATLAB code is shown below:

```matlab
function PhasePlot2
%%%%%%%%%%%%%%%%%%%%%%%%%%%%%%%%%%%%%%%%%
% Title: phase plot
% Author: Marco Albrecht
% Licence: EUPL v1.2
% Last change: 01.09.2018
clear global ;close all ; clc ;
%%%%%%%%%%%%%%%%%%%%%%%%%%%%%%%%%%%%%%%%%
%% Parameters
p.a= 1;%
p.b= 1;%
%% Trajectories
%Trajectory 1
 tspan=0:0.01:150;%[t_start t_end]
    c0=[0.2; 2];%[initial conditions]
    [¬,c1] = ode15s(@system,tspan,c0,[],p);
%Trajectory 2
    tspan=0:0.01:150;%[t_start t_end]
    c0=[2.5;2];%[initial conditions]
    [¬,c2] = ode15s(@system,tspan,c0,[],p);

%% Plot
    figure(1)
    set(gcf,'color','w');
    hold on
    % trajectories
    plot(c1(:,1),c1(:,2),'Color', [0.4 ...
        0.75 0.75],'Linewidth',2)
    plot(c2(:,1),c2(:,2),'Color', [0.85 ...
        0.6 0.85],'Linewidth',2)
    % Asymptotes
    plot([0 2.5],[0 2.5],'k--')
    plot([0 2.5],[1 1],'k--')
    % steady states
    plot(0,0,'ro','MarkerSize',3,'LineWidth',5)
    plot(1,1,'ro','MarkerSize',3,'LineWidth',5)
    % arrows
    annotation('arrow',[0.54/2.5 ...
        0.52/2.5],[1.8/2.5 ...
        1.6/2.5],'Color', [0.4 0.75 ...
        0.75],'Linewidth',2)
    annotation('arrow',[1.2/2.5 ...
        1.3/2.5],[0.8/2.5 ...
        0.87/2.5],'Color', [0.4 0.75 ...
        0.75],'Linewidth',2)
    annotation('arrow',[1.5/2.5 ...
        1.3/2.5],[1.9/2.5 ...
        1.78/2.5],'Color', [0.85 0.6 ...
        0.85],'Linewidth',2)
    % etc
    ylabel('Y')
    xlabel('X')
end
function dcdt=system(¬,c,p)
%% Load parameters
% p.a=1
% p.b=1
%% System dxdt=A*x
dcdt(1)  =    p.a*c(1)*(1-c(2))      ;% x
dcdt(2)  =    p.b*c(2)*(c(1)-c(2))   ;% y
dcdt = dcdt';   % important! - transposes ...
        the solution vector
end
```

Notes

Notes

Chapter 4: Physical modeling and non-linear enzyme kinetics

Thomas Sauter, Marco Albrecht

Motivation

Reality can be difficult to grasp. Philosophers puzzle their heads over the true nature of reality and our image of it. Accordingly, modeling is the art of coming close enough to reality to give valuable insights, without actually being reality. Sometimes we have to include more details. The reaction rates v, with which you are already familiar, can be replaced by more complicated non-linear enzyme kinetics under certain circumstances. Moreover, mathematical equations might be too flexible initially because they are entirely virtual. We have to actively include physical constraints to let the model behave more like reality and to force the model behavior into realistic paths. Different modeling techniques can require the application of different physical laws. But instead of making our models more and more complicated, we have to ask ourselves: what is essential and thus has to be included in a model and what only distracts us from the real kernel? But what is the real kernel? Which criterion can help us to choose an appropriate approximation of our problem? We can neither ignore reality nor can we ever fully represent it. Join us in finding a healthy distance from reality and in exploring the fascinating and comprehensive field of enzyme kinetics. Modeling is a wonderful tool for thinking deeply about the origin of all these superficial observations which we encounter in private life and our professional existence. Let's make this tool sharp!

Keywords

Model classification — Model building — Assumptions — SI units — Akaike information criterion — Balance equation — Enzyme kinetics

Contact: thomas.sauter@uni.lu. **Licence:** CC BY-NC

Contents

1. Lecture summary

We learned in the previous chapters how to set up system equations and understand their dynamics. This is especially helpful for Metabolic Network Analysis. There, already, the graph alone can give us insight. However, we have not modeled physical systems. In order to model physical systems, we need to think about physical laws in thermodynamics, include physical relevant parameters, think about reaction laws, and get familiar with unit calculus. However, we begin with the question of what distinguishes reality from its approximation.

If you prefer, you could also jump directly to balancing (page 10)—an approach which enables us to formulate mathematical models of physical and biological systems.

1.1 ■ What is a model?

"Un cercle n'est pas absurde,[..] Mais aussi un cercle n'existe pas" / "A circle is not absurd; it is clearly explained by the rotation of a straight segment around one of its extremities. But neither does a circle exist." This citation is from the work *La Nausée* written by Jean-Paul Sartre[1] [1, 2]. Sartre was desperate because whenever he had a word for or a model of something, it would never be the reality he wanted to describe. Because we are unable to name something appropriately, the real things remain nameless. The difference between the beauty of abstract description and the imperfection of real-world

---

[1] French philosopher, novelist, political activist, playwright, literary critic, and biographer Jean-Paul Charles Aymard Sartre (1905—1980).

phenomena nauseated him because it revealed to him the uselessness of existence. In the world of mathematics, everything seems to be perfect and pure—like a circle. But you will never find a perfect circle in reality. Thus, the variety of human existence might suggest that something perfect—or a higher power which leads us—cannot exist. Instead, it is up to us to turn our existence into essence. Independent of our opinion of the European movement of existentialism, this movement urges us to get a balanced feeling for what models are. We have to bring them close enough to reality to be useful, but one has to stop the refinement if this stops helping us answering real-world questions. George Box[2] once wrote: "Remember that all models are wrong; the practical question is how wrong do they have to be to not be useful" [3]. This pragmatic view helps us navigate the process of understanding reality with useful models, but it spares us depression caused by exaggerated perfectionism. Mechanistic models have the following advantages [3]:

- They contribute to our scientific understanding of the phenomena under study.

- They usually provide a better basis for extrapolation (at least to conditions worthy of further experimental investigation, if not through the entire range of all input variables).

- They tend to be parsimonious (*i.e.* frugal) in the use of parameters and provide better estimates of the response.

We want to address what the reality is in our specific case. Biologists know that experimental models in biology should resemble—but usually do not fully represent—real-world phenomena *in vivo*. However, the biologist does not really need to think about characteristics he or she does not care about. We biologists have enough trouble ensuring that one's experimental setting allows one's living system to operate in the desired states and that the observations (measurements) indeed show what the real system states look like in principle. It is also the pragmatic realization that it is just impossible to measure all possible players at any time under any condition. But, many things in biologists' system of interest are already true because they work on a real system. If they model something, they work on an entirely virtual system which they think represents what they care about. This does not mean it is true. It is an abstraction. One has to actively ensure that one's mathematical model is physically correct, to figure out which elements one can neglect and which not, and to iteratively compare one's simulation result with experimental data. Which constraints one must take into account depends on one's chosen modeling framework, its inherent mathematical

characteristics, and the underlying assumptions. Consequentially, the interpretation of the model's result depends on one's modeling strategy and the considered assumptions, and is restricted to the considered conceptualization. This applies to the modeling framework itself and to the content of the model. Modeling helps us to dive into a deeper understanding of what we observe and to find out what does not work or maybe why our conception of reality fails. Modeling is not a tool to confirm our premature beliefs about what is helpful in order to seem successful. The less we know, the more careful we should be with interpretation. Modeling is a tool to helps us understand biology, but it does not replace experiments. Bottom-up modeling helps us to develop a consistent theory on what we observe and top-down modeling help us to extract the most likely relationships, but modeling cannot magically generate new data from nothing. Never fall in love with your model and always be sceptical and aware of its shortages. Moreover, it is not an excuse to oversimplify a model until it shows the obvious and thus becomes useless. Useful modeling is the art of representing reality to a degree that goes beyond what you can understand on your own and find the limits of what the mathematical framework can handle with given data. Therefore, you need to master the mathematical framework, its physical interpretation, and the biology of what your model represents. You need the knowledge on how experimental data has been derived, how data can be interpreted, and whether you can use this data in the context of your model. The general modeling approach is part of the next section.

1.2 ■ Modeling procedure

A mathematical model uses mathematical descriptions and computer simulations. A model can be defined as:

Definition 1. Model: A model is the representation of the essential aspects of a system which represents knowledge of that system in usable form (Eykhoff, 1974) [4].

The modeling procedure is an iterative approach and follows a cycle such as that shown in Figure 1 and usually requires an iterative procedure of computational and experimental analysis (Figure 2). The next section is based on a textbook in German language [5].

Specify the question of interest
The question of interest is of utmost importance. It will hardly ever be the case that you have a universal model with which you can explain all kind of questions. With the main question in mind, you can start with the simplest model that can roughly represent your system. As soon as it works, you can refine it iteratively. Modeling scopes can be:

- Scientific understanding
- Validation or falsification

---

[2] British statistician George Edward Pelham Box (1919—2013).

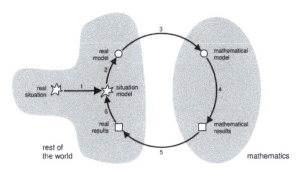

Figure 1. Modeling cycle of Blum and Leiß. 1: understanding **2:** simplifying/structuring **3:** mathematizing **4:** working mathematically **5:** interpreting **6:** validating [6]. Copyright © 2007 Authors (Blum, Leiß), Fair Use.

- Prediction
- Decision-making, strategy planning
- Design of a controller
- Virtual prototype or material simulation
- Pattern recognition.

The modeling scope determines the level of detail of the model. An experienced modeler can plan the modeling strategy with the mathematical implementation and numerical issues in mind. Numerical analysis[3] is a mathematical discipline which tends to solve problems approximately. Thereby, errors can occur which have to be minimized. Not all numerical methods can be applied to any problem, and sometimes the solving process can go completely wrong.

Assumptions
Assumptions about reality help us to make the real-world problem abstract and solvable. There are, for example, assumptions for simplifying the description of inter-dependencies or assumptions for constraining the number of states to investigate. Assumptions should not contain the obvious, should enable a useful model, and should not lead to infeasible models. Assumptions build the foundation of the model and determine the interpretation of the outcome. Each assumption has to be justified and checked for the qualitative and quantitative impact on the model. The assumptions should not contradict one another. Assumptions might change during the modeling process. The following questions help us to identify appropriate assumptions:

- Which effects can be neglected?
- Which scenario might be a good starting case?
- Which parts of a system are necessary and which are not relevant?
- Can a part of the system be prescribed?

---

[3] Earliest mathematical writings from Babylon 1800—1600 BC (Iraq).

Model building
The modeling procedure starts with a **verbal model** which is the initial model concept. It helps us to determine what belongs to the model and what does not, and is specified by the expert of the system. Defining the system's purpose is a part of this process. The verbal model is written in the everyday language of the modeler and the experts. It is also the starting point for the influential structure of a given system. The following steps are necessary to build a model in order to answer the question of interest:

- Specify model and system parameters
- Specify model variables and system states
- Specify the interdependencies between the states with auxiliary conditions and physical laws
- Formulate the mathematical procedures, such as optimization, sensitivity analysis etc.

The **system variables** represent the states which change during the simulation. The system variables span the phase space and should fulfil the following criteria:

- **Independence:** No system variable can be described as a function of another.

- **Completeness:** The state of the system is fully explained by all state variables.

The **auxiliary conditions** describe the interdependencies between the system variables and are adjusted by the system parameter. Modelers use concept maps to draw potential cause-effect relationships and to reason about what has to be modeled and how. Engineers are, like biologists, very visual scientists and need drawings for fruitful communication.

A **system parameter** is a number that does not change with time. In the previous chapters, we mainly talked about rates. These rate terms represent equations by themselves and include parameters. Only in the most simple case are rates represented by a single parameter. A typical procedure in modeling is the search for parameters which are able to let the model behave comparably to reality.

A good procedure to get good models is to ensure that one has as many equations as variables (remember the Rouché–Capelli theorem).

1.3 ■ Mathematical analysis of models
Mathematical methods can be used to make models feasible, to solve them and to understand their properties. This section is based on the same textbook in German language [5].

Dimension analysis
The model variables might have physical meaning and units. A reformulation can be helpful to get rid of the units with a proper unit calculus. This process is called

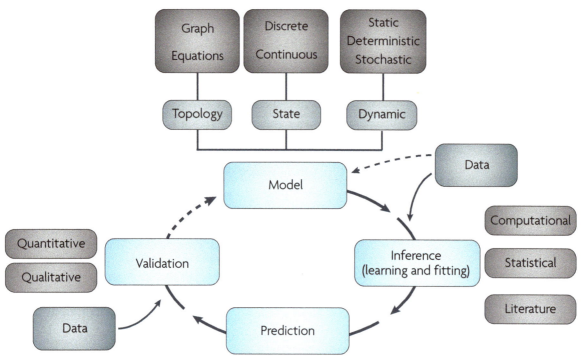

Figure 2. The principle of iteration in modeling. We usually start with a hypothesis and choose an appropriate modeling strategy (graph, discrete, continuous, static, stochastic, deterministic). We integrate data and adjust the model until it fits. Then we perform a prediction and see whether the model is qualitatively and quantitatively good enough. If not, the cycle is repeated until we sufficiently understand the system. More information and source: [7]. Copyright © 2009, Nature Publishing Group.

non-dimensionalization. Moreover, relevant lengths or magnitude ranges of the variable values might be defined to determine what relevant influences are. Questions of scale are relevant. Multi-scale models can contain different relevant magnitude orders of interest.

Inverse and poorly posed problems

One might perform certain steps to check whether the model is mathematically well-posed in itself. If not, mathematical procedures might perform poorly. A well-posed problem has a unique solution. This solution changes continuously with continuously changing parameters. A system matrix which is not invertible can cause numerical problems. Matrices with a high condition number are ill-conditioned. A condition number measures how much an output value changes with small changes of the input number.

Investigate special problems

One might first solve special cases before investigating the more complicated general case.

Simplification and linearization

One might neglect terms with minor impact to see if the simplified model behaves similarly to the original model. The neglected terms are called small disturbances of the

simplified model (perturbation theory). Karl Popper [4] once said: "Science may be described as the art of systematic over-simplification—the art of discerning what we may with advantage omit."

Linearization around specific points might help investigate the model's behaviour and acquire information about the non-linear model at specific points.

Check qualitative properties

One runs the model and gets conclusions on asymptotic behavior, system stability, and if oscillations can occur, among other qualitative properties.

Sensitivity analysis

After one has found a solution with a given set of optimal parameters, one wants to know how small changes of parameter values influence the solution. This is an important type of analysis to carry out.

Interpretation and validation

The interpretation of the results might happen in light of previous expectations and requires the consideration of all assumptions. The simulation might give us a prediction which is then subjected to a validation process. If experimental data can be generated, one has to determine how tolerant one is ready to be. How well does

---

[4] Austrian and British philosopher Sir Karl Raimund Popper (1902—1994).

the model fit the new data? Sometimes no data can be generated. In this case, check the plausibility with the experts of the given system. Surprising simulation results might lead to a deeper investigation of subsystems or checking whether parameters and variables remained in physically meaningful ranges after the previous optimization. One cannot prove the 'correctness' of a model, and a model cannot be 'verified'. Correct settings in a simulation cannot constitute proof of accurate behavior under all possible circumstances. The only thing it can do is falsify the theories and models, *e.g.* if observations and simulations disagree. Consequently, we always say that we validate a model in light of the model's purpose and in the following aspects:

Structural validity: The model has an influence structure which resembles the influence structure of the original. The number of essential states and the feedback structures have counterparts in the original system.

Behavioral validity: The model shows qualitatively the same dynamic behavior, *e.g.*, oscillations or not.

Empirical validity: Within the model's constraints, the simulation result follows experimental data and, if no observable data is available, the results are at least consistent and plausible, and the parameter ranges are empirically justified.

Application validity: The model fulfils its purpose and can be applied to make decisions, to control processes etc.

1.4 ■ Model classification

Usually you do not have to reinvent a modeling strategy. Often you can use previous procedures and apply these to your problem of interest. How often which method type is used can be seen in Figure 3. The spectrum of some methods in systems biology can be seen in Figure 4. Mathematical approaches are listed in Figure 16 in the section on mathematical basics. Here, we give an overview on model types. This section is based on the same textbook [5].

Mathematical structure

We can distinguish the following contrasting pairs:

Static or dynamic models: Dynamic models change over time. Static models are time-independent or are in the steady state of a dynamic model. The time can be continuous or discrete.

Discrete or continuous models: If the variable values can be counted, one is dealing with a discrete model. See Figure 5 for an illustration of continuous and discrete signals in time and state. Each class requires fundamentally different modeling strategies. Discrete models require methods of graph theory and the theory of finite-state machines. Continuous models require higher mathematical models of calculus. However, it is possible to discretize continuous models to obtain and handle a simpler and reduced discrete model in the hope it sufficiently resembles the original continuous model.

Deterministic or stochastic models: A deterministic model always comes to the same solution with given parameters and initial variable values. Stochastic models operate with random events and are used for models with certain likelihoods. Stochastic models are also used for very complex models where the structure is not fully understood.

Sometimes we have models with mixed elements; however, it helps to understand the character of each in order to make wise choices and to be aware of the consequences.

Level of description

1st-principle models, white-box models   Models can be assembled from known laws or well-understood 1st principles.

Heuristic models, gray-box models   Models based on more or less justified assumptions for the cause and effect inter-relationships. Often for complicated systems without stringent rules, such as ecologic systems and the Predator-Prey Model.

Descriptive models, black-box models   Models generated on the basis of data without system knowledge. Typically input-output relationships, as illustrated in Figure 6.

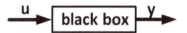

Figure 6. The black-box concept. We know little about the system, but we can learn from the input u - output y behavior.

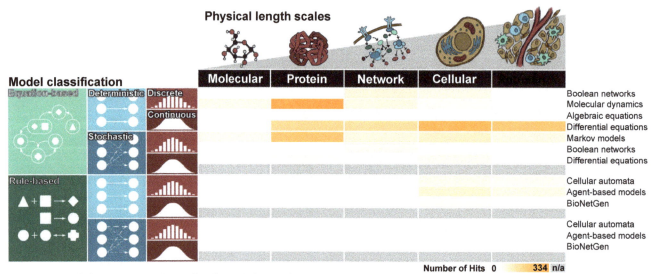

Figure 3. **Model types:** Number of PubMed 'hits' of a model class for a given biological problem. Adapted from [8]. Copyright © 2016, Elsevier Ltd.

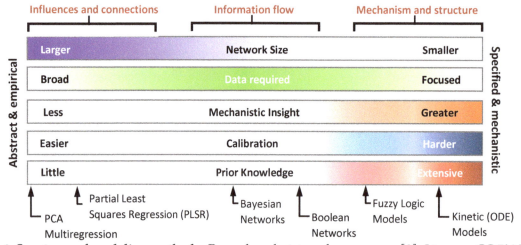

Figure 4. **Spectrum of modeling methods:** Reproduced picture from source: [9]. Licence: CC BY-SA 4.0

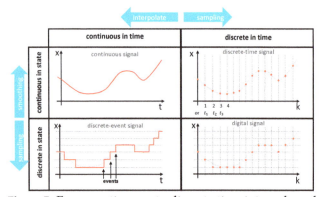

Figure 5. From continuous to discrete time intervals and discrete states.

1.5 ■ Akaike Information Criterion

"Whenever a theory appears to you as the only possible one, take this as a sign that you have neither understood the theory nor the problem which it was intended to solve," said Karl Popper. Therefore, we usually have several candidate models and ask ourselves which model might be the best. A scheme is shown in Figure 7.

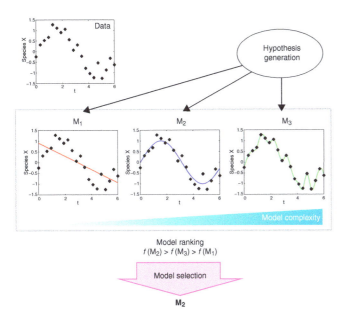

Figure 7. Model selection: We generate different hypotheses and select the best-balanced model. Model M_1 is too simple to mimic reality (not enough parameters). Model M3 fits the noise rather than the underlying system and is thus not robust (too many parameters, overfitting). Model M2 represents a good fit. Source: [10]. Copyright © 2013, Elsevier Ltd.

A commonly used decision tool for the model selection is the Akaike Information Criterion (AIC), amongst other similar criteria. The AIC takes into consideration the goodness of fit $\hat{L}$ and the number of parameters k:

$$AIC_i = 2k - 2\ln(\hat{L}).$$

The model with the minimal AIC should be preferred. This criterion says nothing about whether a model is good or bad. It simply says that a model is capable of generating behavior in line with data. The relative likelihood of a subordinate model i to the minimal model AIC$_i$ can be calculated with:

$$\exp((AIC_{min} - AIC_i)/2)$$

For small models, the AIC might point to models with too many parameters. The model is overfitting then. It means that you fit noise instead of the model dynamic, which increases the likelihood of giving false predictions.

With the following correction:

$$AICc = AIC + \frac{2k^2 + 2k}{n - k - 1}$$

you increase the impact of the parameters. A large sample size n can reduce the impact of this correction factor. For those of you who want to learn more, there is also the Bayesian Information Criterion (BIC), the Takeu-chi's Information Criterion (TIC), and the Widely Applicable Information Criterion (WAIC) [10]. The model selection can thereby depend on the experimental design [11]. Minimizing the AIC is effectively equivalent to maximizing entropy in a thermodynamic system according to the second law of thermodynamics.

1.6 ■ Extensive properties & SI units

Besides continuous macro-scale and stochastic micro-scale thermodynamics, we have to differentiate between extensive quantities[5] and intensive quantities, if it comes to transporting from one system to another. Intensive quantities such as temperature, concentration, pressure, and density are independent of the system size or amount of material. Thus they cannot be physically transported through a system boundary. Extensive quantities, however, are defined quantities whose magnitude is additive for subsystems, according to the International Union of Pure and Applied Chemistry IUPEC. Around 30 extensive quantities are listed in Table 1 and are grouped according to their relationship to mass.

The old SI units (International System of Units) in Figure 8 are currently under reformulation. We mention the newest proposals:

Definition 2. Second: The second, symbol s, is the SI unit of time. It is defined by taking the fixed numerical value of the caesium frequency ΔvCs, the unperturbed ground-state hyperfine transition frequency of the caesium 133 atom, to be 9192631770 when expressed in the unit Hz[6], which is equal to s^{-1}.

Definition 3. Meter: The meter, symbol m, is the SI unit of length. It is defined by taking the fixed numerical value of the speed of light in vacuum c to be 299792458 when expressed in the unit $m \cdot s^{-1}$, where the second is defined in terms of the caesium frequency ΔvCs.

Definition 4. Kilogram: The kilogram, symbol kg, is the SI unit of mass. It is defined by taking the fixed numerical value of the Planck[7] constant h to be $6.62607015 \cdot 10^{-34}$

---

[5] Most textbooks use the term "extensive properties"; however, this term is impure in this context because properties like color are not quantitative, and some extensive quantities such as heat and work are not properties [12].

[6] German physicist Heinrich Rudolf Hertz (1857—1894).

[7] German theoretical physicist Max Karl Ernst Ludwig Planck (1858—1947).

Table 1. Extensive properties: This table gives you a feeling for different extensive properties. You do not need to learn this by heart—you simply need to understand the principle. Z: extensive property. m: mass. z: intensive property.

Symbol	Quantity	Unit	Conditions	Formula	Notes
				$(Z/m =)$	
Z1			*proportional to mass*		
m	mass	kg	none[a]	$m/m = 1$	
F_g[b]	weight	J m^{-1}	g constant	$F_g/m = g$	$g = 9.81$ m s^{-2}
Z2			*proportional to mass under constant composition $(Z/m = k)$, else Z2 = Z3,*[c]		
C_p	heat capacity	J K^{-1}	n_i constant	$C_p/m = c_p$	c_p = specif c heat cap.
n	amount of substance	mol	n_i constant	$n/m = 1/M$	M = molar mass
V_{SL}	volume (solid, liquid)	m^3	n_i constant	$V_{SL}/m = \rho$	ρ = density
N	number of particles	–	n_i constant		
Z3			*mass is constant of proportionality, or conditional proportional $(Z/m = z)$*		
E_k	kinetic energy	J		$E_k/m = \frac{1}{2}v^2$	v = velocity
E_g	gravitational energy	J		$E_g/m = gh$	h = height, g = 9.81m/s^2
p	momentum	kg · m s^{-1}		$p/m = v$	v = velocity
F[b]	force	J m^{-1}		$F/m = a$	a = accelaration
U	internal energy	J		$U/m = u(T)$	T = temperature
G	free enthalpy	J		$G/m = g(T,p)$	g = specif c free enthalpy
H	enthalpy	J		$H/m = h$	h = specif c enthalpy
S	entropy	J K^{-1}		$S/m = s$	s = specif c entropy
V_G	volume (gas)	m^3		$V_G/m = M.v_G(T,p)$	v_G = molar volume
V	volume (all phases)	m^3		$V = V_{SL} + V_G$	
Z4			*extensive quantities independent of mass*		
A	area (interface)	m^2		–	
E_{el}	electric energy	J		–	
E_{sp}	spring energy	J		$E_{sp} = \frac{1}{2}k.x^2$	k = spring constant
E_{pv}	displacement energy	J		$E_{pv} = p.V$	p = ambient pressure
E_{sur}	surface energy	J		$E_{sur} = \gamma.A$	γ = surface tension
i	current (electric-)	C s^{-1}		–	
P	power (most forms)	J s^{-1}		–	–
Q	heat	J		–	–
q	charge (electric)	C		–	
W	work	J		–	–
ξ	extent of reaction	eq[d]		–	conjugate to aff nity
$d\xi/dt$	reaction rate	eq s^{-1} [d]		–	
	f ow (f uid)	kg s^{-1}		–	
	total quantities = 29	total			

when expressed in the unit J · s, which is equal to kg · m^2 · s^{-1}, where the meter and the second are defined in terms of c and $\Delta\nu Cs$.

Definition 5. Ampere: The ampere,[8] symbol A, is the SI unit of electric current. It is defined by taking the fixed numerical value of the elementary charge e to be $1.602176634 \cdot 10^{-19}$ when expressed in the unit C, which is equal to A · s, where the second is defined in terms of $\Delta\nu Cs$.

[8] French physicist and mathematician André-Marie Ampère (1775—1836).

Old SI

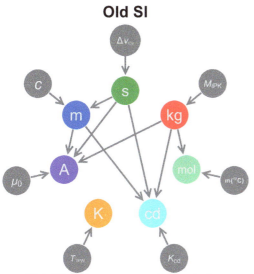

New SI

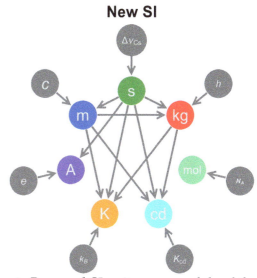

Figure 8. Old SI units: s: second, kg: kilogram, mol: mole, cd: candela, K: kelvin, A: ampere. m: meter. Source: Wikipedia, Emilio Pisanty, 2016, Licence: CC BY-SA 4.0.

Figure 9. Proposed SI units: s: second, kg: kilogram, mol: mole, cd: candela, K: kelvin, A: ampere. m: meter. Source: Wikipedia, Emilio Pisanty, 2016, Licence: CC BY-SA 4.0.

Definition 6. Mole: The mole, symbol mol, is the SI unit of amount of substance. One mole contains exactly $6.02214076 \cdot 10^{23}$ elementary entities. This number is the fixed numerical value of the Avogadro constant,[9] N_A, when expressed in mol^{-1}, and is called the Avogadro number (IUPEC, 2018, [13]).

Definition 7. Amount of substance: The amount of substance, symbol n, of a system is a measure of the number of specified elementary entities. An elementary entity may be an atom, a molecule, an ion, an electron, or any other particle or specified group of particles (IUPEC, 2018, [13]).

Definition 8. Dalton: The 1971 definition of the mole implies that the Avogadro number equals the ratio of the gram to the Dalton[10] (mu = 1 u = 1 Da), with the value of the Dalton (Da) expressed in grams. The historical continuity of the present definition preserves this relation, not exactly, but to within an uncertainty 10^{-10} negligible for practical purposes (IUPEC, 2018, [13]). The relationship of Dalton to the Avogadro constant might be redefined to a scaling factor $N_A = \frac{g}{Da} \cdot mol^{-1}$.

Definition 9. Kelvin: The kelvin,[11] symbol K, is the SI unit of thermodynamic temperature. It is defined by taking the fixed numerical value of the Boltzmann[12] constant K to be $1.380649 \cdot 10^{-23}$ when expressed in the

unit $J \cdot K^{-1}$, which is equal to $kg \cdot m2 \cdot s^{-2} \cdot K^{-1}$, where the kilogram, meter, and second are defined in terms of h, c and ΔvCs.

Definition 10. Candela: The candela,[13] symbol cd, is the SI unit of luminous intensity in a given direction. It is defined by taking the fixed numerical value of the luminous efficacy of monochromatic radiation of frequency $540 \cdot 10^{12}$ Hz, K_{cd}, to be 683 when expressed in the unit $lm \cdot W^{-1}$, which is equal to[14] $cd \cdot sr \cdot W^{-1}$, or $cd \cdot sr \cdot kg^{-1} \cdot m^{-2} \cdot s^{3}$, where the kilogram, meter, and second are defined in terms of h, c and ΔvCs.

With this new SI set, the relationships between the SI units might change according to the scheme in Figure 9.

---

[9] Italian scientist Amedeo Carlo Avogadro (1776—1856).

[10] English chemist, physicist, and meteorologist John Dalton (1766—1844).

[11] Scots-Irish mathematical physicist and engineer William Thomson Kelvin (1824—1907).

[12] Austrian physicist and philosopher Ludwig Eduard Boltzmann (1844—1906).

[13] From candle.

[14] sr: The steradian or square radian is the SI unit of solid angle.

1.7 ▨ Balancing

After this detailed introduction to modeling concepts and other relevant background information, we are now coming back to the question on how to formulate mathematical models of biological systems. We will introduce the approach of balancing here, which is an extension of the ODE equations derived for biochemical reactions in the previous chapters. The general concept of balancing for describing system behavior over time is:

$$\boxed{\text{temporal change of state variables}} = + \boxed{\text{inflow}} - \boxed{\text{outflow}} + \boxed{\text{source}} - \boxed{\text{sink}}$$

Inflows and **outflows** are exchanges across the system boundary. In contrast, **sources** and **sinks** are system internal processes.

Continuous changes are described with differential equations.

$$\frac{d\Phi}{dt} = \dot{\Phi} = \underbrace{J + P}_{\text{Rates of change}}$$

In a (time) discrete form this would be:

$$\Delta\Phi = \Phi(t+1) - \Phi(t) = \underbrace{J + P}_{\text{Changes}}$$

$$\Rightarrow \quad \Phi(t+1) = \Phi(t) + J + P$$

Example 1: Population balance

A continuous population balance with population size N $[-]$.

$$\boxed{\dot{N}} = + \boxed{\begin{array}{c}\text{immigration}\\\text{rate}\end{array}} - \boxed{\begin{array}{c}\text{emigration}\\\text{rate}\end{array}} + \boxed{\begin{array}{c}\text{birth}\\\text{rate}\end{array}} - \boxed{\begin{array}{c}\text{death}\\\text{rate}\end{array}}$$

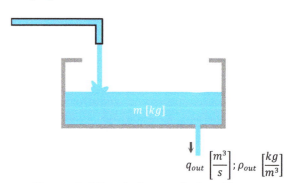

$$q_{in}\left[\frac{m^3}{s}\right]; \rho_{in}\left[\frac{kg}{m^3}\right]$$

$$m\,[kg]$$

$$q_{out}\left[\frac{m^3}{s}\right]; \rho_{out}\left[\frac{kg}{m^3}\right]$$

Figure 10. Mass balance of a liquid container.

Example 2: Mass balance

The mass balance of a liquid container is as shown in Figure 10:

$$\dot{m} = q_{in}\cdot\rho_{in} - q_{ex}\cdot\rho_{ex} \qquad \left[\frac{kg}{s}\right]$$

Example 3: Volume balance

The volume balance can be derived from the mass balance:

$$\dot{m} = (\rho\overset{\cdot}{\cdot}V) = \rho\cdot\dot{V} + \dot{\rho}\cdot V = \rho\cdot\dot{V}$$

with the product rule, and with the assumption that the density is constant and does not change $\rho = \text{const}$.

$$\dot{\rho} = 0$$

Example 4: Amount of substance balance

The amount of substance balance is a type of mass balance which is especially suited for chemical reactions. It is also known as mole balance. The change of the amount of substance is $\dot{n}$ with the unit $\left[\frac{mol}{s}\right]$. The rate constant is k and can have different units. Let us look at the following reaction:

$$A + B \xrightarrow{k} P$$

with the balance equation:

$$-\dot{n}_A = -\dot{n}_B = \dot{n}_P = k\cdot C_A\cdot C_B\cdot V$$

The unit calculus is

$$\left[\frac{mol}{s}\right] = X\cdot\left[\frac{mol}{l}\right]\cdot\left[\frac{mol}{l}\right]\cdot 1$$

The missing unit X for the rate constant k is therefore $\left[\frac{1}{s \cdot mol}\right]$. Remember: $n_A = C_A \cdot V$

$$\Rightarrow \dot{n}_A = (C_A V)^{\cdot} = \dot{C}_A V + C_A \dot{V} = \dot{C}_A V$$

with the product rule and constant volume $V = $ const:

$$\dot{V} = 0$$

We moreover obtain the following relationships:

$$\Rightarrow -\dot{n}_A = -\dot{C}_A \cdot V = k \cdot C_A \cdot C_B \cdot V$$

$$\Rightarrow \qquad \dot{C}_A = -k \cdot C_A \cdot C_B$$

We might also point to the rate of mole change:

$$\dot{n}_i = q \cdot c_i$$
$$\left[\frac{mol}{s}\right] = \left[\frac{m^3}{s}\right] \cdot \left[\frac{mol}{m^3}\right]$$

$$A + B \underset{k_{-1}}{\overset{k_1}{\rightleftharpoons}} P \qquad \frac{dn_A}{dt} = -k_1 \cdot c_A \cdot c_B \cdot V + k_{-1} \cdot c_P \cdot V$$

$$A + B \underset{k_{-1}}{\overset{k_1}{\rightleftharpoons}} P + S \qquad \frac{dn_A}{dt} = -k_1 \cdot c_A \cdot c_B \cdot V + k_{-1} \cdot c_P \cdot c_S \cdot V$$

Some useful equations
The **molar mass** is:

$$M_i = \frac{m_i}{n_i} = \frac{\rho_i}{c_i} \quad \rightsquigarrow \quad m_i = M_i \cdot n_i$$

The **mass flow** $\dot{m}_i \left[\frac{kg}{s}\right]$ is:

$$\dot{m}_i = q \cdot \rho_i$$

with volume flow rate $q \left[\frac{m^3}{s}\right]$ and mass density $\rho \left[\frac{kg}{m^3}\right]$.

The **molar flow** $\dot{n}_i \left[\frac{mol}{s}\right]$ is:

$$\dot{n}_i = q \cdot c_i$$

with volume flow rate $q \left[\frac{m^3}{s}\right]$ and molar concentration $c_i \left[\frac{mol}{m^3}\right]$. The conversion is:

$$\frac{mol}{m^3} = 10^{-3} \frac{mol}{L} = 10^{-3} M = 1 mM$$

with the unit molar M and millimolar mM.

Reactions
We repeat some reactions and their respective balance equations:

$$A \xrightarrow{k_1} P \qquad -\frac{dn_A}{dt} = \frac{dn_P}{dt} \qquad \frac{dn_A}{dt} = -k \cdot c_A \cdot V$$

$$A \underset{k_{-1}}{\overset{k_1}{\rightleftharpoons}} P \qquad \frac{dn_A}{dt} = -k_1 \cdot c_A \cdot V + k_{-1} \cdot c_P \cdot V$$

1.8 ■ Enzyme kinetics

Enzymes are very important proteins and good drug targets. For example, the BRAF kinase[15] can be inhibited by the drug dabrafenib. Dabrafenib in turn can be degraded by another enzyme class named cytochrome P450. The CYP450 enzyme class accounts for 75% of drug metabolism and five of the 57 human CYPP450s are involved in 95% of these reactions [14]. Enzymes are sometimes very unspecific and share similar targets. Thus when an enzyme is inhibited, this can result in undesired off-target effects also on other enzymes. Have a look at the phylogenetic tree of the human kinome in Figure 11.

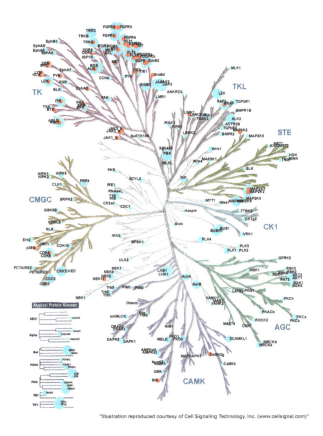

"Illustration reproduced courtesy of Cell Signal Technology, Inc. (www.cellsignal.com)"

Figure 11. Kinase tree: The maps shows the phylogenetic tree of the human kinome using the KinMap software. Sky-blue annotations are kinases which are melanoma-specific. Red dots indicate kinases for which drugs are available. Source: [15], Licence: CC BY-SA 4.0.

Chemoinformatics can analyze the molecular structure of both enzyme pockets and drugs together. 3D pockets of key targets and off-target kinases thereby allow the design of more specific drugs [16]. This helps us to predict off-target effects, enzyme-drug affinities, and support the improvement of drug structures. We will

concentrate on the enzyme kinetic for biochemical networks. In Systems Biology, we are often more interested in the interplay of an enzyme with other molecules and the role of an enzyme in a network. We might thereby ignore the chemical complexity as well as the structural geometry of the enzymes and ask *e.g.* only the following questions:

- Which molecules can bind the enzyme?
- Which molecules will be released?
- Which molecule-enzyme affinities appear?
- Is the binding reversible?
- Does the molecule inhibit or activate the enzyme function?
- Does the molecule bind the catalytic side or an allosteric side?
- Is the enzyme specific?

YouTube: Enzyme kinetics series (7 videos by Khan Academy)

The reaction from substrate to product is usually promoted by enzymes. Enzymes split the reaction into less energy-intensive subreactions and thereby have a binding phase and a catalytic phase, as shown in Figure 12. You learned in Chapter 2 that we can model biochemical networks with the stoichiometric matrix:

$$\dot{S} = N \cdot v$$

which we often scrutinize in the steady state (dynamic equilibrium) with:

$$0 = N \cdot v$$

The non-trivial solution gave us the fluxes and interesting properties such as dead ends, linear paths, and the kernel matrix K. We repeat the balancing of a biochemical reaction in Example 5. In this section we will focus on the reaction vector v.

---

[15] Kinases catalyses the transfer of the phosphate group to specific substrates. The opponent is the phosphatase.

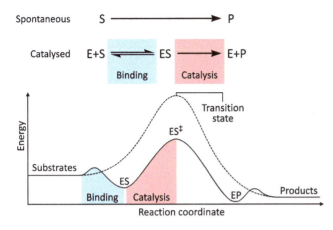

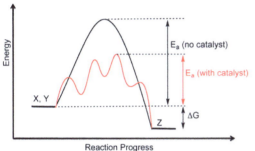

Figure 12. Energy profile during an enzymatic reaction: The bigger the energy difference, the lower the probability of transformation to the products. Consequently, the reaction rate is lower as well. Source: Thomas Shafee & Smokefoot, Wikimedia, Licence: CC BY 4.0 (upper figure) & CC0 1.0 Universal Public Domain Dedication (lower).

Example 5: Balancing a biochemical reaction

The substrates are catalyzed to form 2 product molecules:

$$S_1 + S_2 \underset{v_b}{\overset{v_f}{\rightleftharpoons}} 2P$$

The overall (net) reaction rate is calculated by the forward reaction rate v_f minus the backward reaction rate v_b. The ODE equation set is:

$$\dot{S}_1 = \dot{S}_2 = -\underbrace{k_1 \cdot S_1 \cdot S_2}_{v_f} + \underbrace{k_{-1} \cdot P^2}_{v_b}$$

$$\dot{P} = 2\underbrace{k_1 \cdot S_1 \cdot S_2}_{v_f} - 2\underbrace{k_{-1} \cdot P^2}_{v_b}$$

The reaction coefficients v_f and v_b are based on the **Law of Mass Action** formulated by Waage[16] and Guldberg[17] in 1864 [17]. We say that the **reaction rate** is proportional to the **probability of a collision** of the reactants. The probability of a collision in turn is proportional to

---

[16] Norwegian chemist Peter Waage (1833—1900).
[17] Norwegian mathematician and chemist Cato Maximilian Guldberg (1836—1902).

---

the concentration of reactants to the power of molecularity (stoichiometric coefficient). Refer once again to Example 5. The **net rate** of the product is:

$$v = v_f - v_b = k_1 \cdot S_1 \cdot S_2 - k_{-1} \cdot P^2$$

with the related unit calculus:

$$\underbrace{\left[\frac{mol}{L \cdot s}\right]}_{v} = \underbrace{\left[\frac{L}{mol \cdot s}\right]}_{k_1} \underbrace{\left[\frac{mol}{L}\right]}_{S_1} \underbrace{\left[\frac{mol}{L}\right]}_{S_2} - \underbrace{\left[\frac{L}{mol \cdot s}\right]}_{k_{-1}} \underbrace{\left[\frac{mol}{L}\right]^2}_{P^2}$$

The **unit of the kinetic parameters** k_i, in the reaction rate v, depends on the number of reaction partners. This unit changes from reaction to reaction.

The general mass action rate law for a single reaction is:

$$v = v_f - v_b = k_{+1} \prod_i S_i^{m_i} - k_{-1} \prod_j P_j^{m_j}$$

m_i ... molecularities of substrates

m_j ... molecularities of products

Reactions in equilibrium $\hat{=}$ concentrations in equilibrium (S_{eq}, P_{eq}).

$$v_f = v_b$$

$$\Rightarrow K_{ea} = \frac{k_{+1}}{k_{-1}} = \frac{\prod_j P_{eq}^{m_j}}{\prod_i S_{eq}^{m_i}}$$

The free energy difference of a reaction:

$$\Delta G = -RT \ln K_{eq}$$

depends on the temperature T and the gas constant $R = 8.314 \frac{J}{mol \cdot K}$. The free energy is indicated in Figure 12.

The Law of Mass Action might result in a lot of equations, and simplifications are often used. These are based on certain assumptions, such as those illustrated in the following for the kinetic rate law of, *e.g.*, Michaelis-Menten. One standard textbook on enzyme kinetics is from Cornish-Bowden[18] [18]. We will derive some of the kinetics and will provide, at the end a summary in Table 2.

Michaelis-Menten kinetics
The Michaelis-Menten kinetic [19] is based on the most common known enzyme reaction scheme:

$$E + S \underset{k_{-1}}{\overset{k_1}{\rightleftharpoons}} ES \xrightarrow{k_2} E + P$$

An enzyme E binds a substrate S to form an enzyme substrate product ES. This enzyme substrate complex can break down either back to the substrate or into the product. In both cases the enzyme is recovered, as expected

---

[18] British biochemist Athelstan John Cornish-Bowden (1943—today).

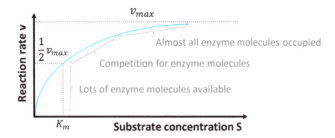

Figure 13. Michaelis-Menten kinetic.

for a catalyst. The balance equations of the reaction network are:

$$\frac{dS}{dt} = -k_1 \cdot E \cdot S + k_{-1} \cdot ES$$

$$\frac{dES}{dt} = k_1 \cdot E \cdot S - (k_{-1} + k_2) \cdot ES$$

$$\frac{dE}{dt} = -k_1 \cdot E \cdot S + (k_{-1} + k_2) \cdot ES$$

$$\frac{dP}{dt} = k_2 \cdot ES$$

The overall reaction rate is:

$$v = -\frac{dS}{dt} = \frac{dP}{dt}$$

The substrate decrease is equal to the product increase. The enzyme is recovered. Because our equation set cannot be solved analytically, we will make some assumptions:

1.) Quasi-equilibrium assumption

Michaelis[19] and Menten[20] stated that the forward and backward reactions between the substrate and the enzyme substrate complex are much faster than the final reaction to the product:

$$k_1, k_{-1} \gg k_2$$

2.) Quasi-steady-state assumption

Briggs[21] and Haldane[22] stated that if and only if the substrate is in much higher abundance than the enzyme $S(t = 0) \gg E$, then the ES complex has constant concentration levels, or more generally:

$$\frac{dES}{dt} = 0$$

with which:

$$\frac{dE}{dt} = 0$$

is obtained. But if neither enzyme levels nor the levels of enzyme substrate change, there must be a total amount of enzyme which we can consider as a constant:

$$E_{total} = E + ES \qquad \Rightarrow \qquad E = E_{total} - ES$$

Putting everything together

Reformulation of the differential equation of the enzyme substrate complex gives:

$$\frac{dES}{dt} = k_1 \cdot E \cdot S - (k_{-1} + k_2) \cdot ES \stackrel{!}{=} 0$$

$$k_1 \cdot (E_{total} - ES^*) \cdot S - (k_{-1} + k_2) \cdot ES^* = 0$$

$$k_1 \cdot E_{total} \cdot S - k_1 \cdot ES^* \cdot S - (k_{-1} + k_2) \cdot ES^* = 0$$

and:

$$ES^* = \frac{k_1 \cdot E_{total} \cdot S}{k_1 \cdot S + (k_{-1} + k_2)} = \frac{E_{total} \cdot S}{S + \frac{k_{-1} + k_2}{k_1}}$$

We are interested in the generation rate of the product depending on the substrate availability:

$$v = \frac{dP}{dt} = k_2 \cdot ES^*$$

which is finally:

$$v = \frac{k_2 \cdot E_{total} \cdot S}{\frac{k_{-1} + k_2}{k_1} + S} = \frac{v_{max} \cdot S}{K_m + S}$$

with the Michaelis-Menten parameter K_m and the maximal catalysis velocity v_{max}. The Michaelis-Menten parameter indicates the substrate concentration with which one can obtain the half-maximal reaction velocity. You will need to be able to distinguish between the parameter types!

Mechanistic parameters: $k_1, k_{-1}, \ldots$ (based on mechanisms)

$$\Leftrightarrow$$

Phenomenological parameters: $v_{max}, K_m, \ldots$ (based on systemic behavior)

Lineweaver–Burk plot

Lineweaver[23] and Burk[24] established a linear representation of the function in Figure 14 to more easily access the parameters. This was of great importance before the appearance of appropriate computer programs. How-

---

[19] German biochemist, physical chemist, and physician Leonor Michaelis (1875—1949).

[20] Female Canadian physician and scientist Maud Leonora Menten (1879—1960).

[21] English scientist John Burdon Sanderson Haldane (1892—1964).

[22] English professor for Botany George Edward Briggs (1893—1985).

[23] American physical chemist Hans Lineweaver (1907—2009).

[24] American biochemist Dean Burk (1904—1988).

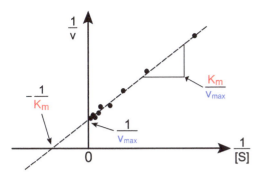

Figure 14. Lineweaver–Burk plot. Source: Wikimedia, Licence: CC BY SA 3.0.

ever, modern methods are available to experimentally determine more physical parameters without the quasi-steady-state assumption [20]. The research is still on-going (for example, taking substrate inhibition into account [21]). Unfortunately, many publications use the kinetic laws although their underlying assumptions are violated—*e.g.* drugs might lead to enzyme induction of the CYP450 class leading to a changing total enzyme concentration E_{total}.

Michaelis-Menten kinetic for reversible reaction
The reaction scheme for the reversible Michalis-Menten kinetic is:

$$E + S \underset{k_{-1}}{\overset{k_1}{\rightleftharpoons}} ES \underset{k_{-2}}{\overset{k_2}{\rightleftharpoons}} E + P$$

where the product concentration changes accordingly:

$$v = \frac{dP}{dt} = k_2 ES - k_{-2} P \cdot E.$$

The rate equation is:

$$v = E_{\text{total}} \frac{q \cdot S - P}{\frac{Sk_1}{k_{-1}k_{-2}} + \frac{1}{k_{-2}} + \frac{k_2}{k_{-1}k_{-2}} + \frac{P}{k_{-1}}} \qquad ; q = \frac{k_1 k_2}{k_{-1} k_{-2}}$$

$$v = \frac{\frac{v_{\text{max}}^f}{K_{\text{mS}}} S - \frac{v_{\text{max}}^b}{K_{\text{mP}}} P}{1 + \frac{S}{K_{\text{mS}}} + \frac{P}{K_{\text{mP}}}}$$

The phenomenological parameters are related in the following way:

$$K_{eq} = \frac{v_{\text{max}}^f K_{\text{mP}}}{v_{\text{max}}^b K_{\text{mS}}}$$

according to Haldene (1930).

General Michaelis-Menten kinetic for inhibition
We look at the general reaction scheme:

$$\begin{array}{ccccc}
E + S & \underset{k_{-1}}{\overset{k_1}{\rightleftharpoons}} & ES & \overset{k_2}{\rightarrow} & E + P \\
+ & & + & & \\
I & & I & & \\
k_{-3} \updownarrow k_3 & & k_{-4} \updownarrow k_4 & & \\
EI + S & \underset{k_{-5}}{\overset{k_5}{\rightleftharpoons}} & ESI & \overset{k_6}{\rightarrow} & E + P + I
\end{array}$$

We distinguish the following cases considering the related reaction numbers 1-6:

1,2 Michaelis-Menten

1,2,3 Competitive inhibition

1,2,4 Uncompetitive inhibition

1-5 Noncompetitive inhibition

1-6 Partial inhibition

We can assume the following binding equilibria between the entities and the complexes:

$$K_m \cong \frac{k_{-1}}{k_1} = \frac{E \cdot S}{ES}$$

$$K_{I,3} = \frac{k_{-3}}{k_3} = \frac{E \cdot I}{EI}$$

$$K_{I,4} = \frac{k_{-4}}{k_4} = \frac{ES \cdot I}{ESI}$$

$$K_{I,5} = \frac{k_{-5}}{k_5} = \frac{EI \cdot S}{ESI}$$

The respective reaction kinetics are listed in Table 2.

Substrate inhibition
The reaction network for substrate inhibition is:

$$\begin{array}{ccc}
E + S & \rightleftharpoons ES & \rightarrow E + P \\
 & + & \\
 & S & \\
 & \updownarrow & \\
 & ESS &
\end{array}$$

which leads to the kinetic:

$$v = k_2 ES = \frac{v_{\text{max}} S}{K_m + S \left(1 + \frac{S}{K_I}\right)} \qquad ; K_I = \frac{k_{-I}}{k_I} = \frac{ES \cdot S}{ESS}$$

with the optimal substrate concentration:

$$S = \sqrt{K_m K_I}$$

and the optimal velocity:

$$v_{\text{opt}} = \frac{v_{\text{max}}}{1 + 2\sqrt{K_m/K_I}}.$$

An example for substrate inhibition of CYP1A2 (CYP450 enzyme class) can be found *e.g.* here [23]. The CYP450

Table 2. Types of inhibition for irreversible and reversible Michaelis-Menten kinetics, so you can see that one needs different kinetic laws for different approaches. It is not necessary to learn this information by heart, but you might ponder on it for a while. Reproduced from source: [22]. Copyright © 2016, John Wiley and Sons.

Name	Implementation	Equation - irreversible	Equation - reversible	Characteristics
Competitive inhibition	I binds only to free E; P-release only from ES $k_{\pm 4} = k_{\pm 5} = k_6 = 0$	$v = \dfrac{v_{max}S}{K_m \cdot i_3 + S}$	$v = \dfrac{v_{max}^f \frac{S}{K_{mS}} - v_{max}^r \frac{P}{K_{mP}}}{\frac{S}{K_{mS}} + \frac{P}{K_{mP}} + i_3}$	K_m changes, v_{max} remains; S and I compete for the binding place; high S may out compete I
Uncompetitive inhibition	I binds only to ES; P-release only from ES $k_{\pm 3} = k_{\pm 5} = k_6 = 0$	$v = \dfrac{v_{max}S}{K_m + S \cdot i_4}$	$v = \dfrac{v_{max}^f \frac{S}{K_{mS}} - v_{max}^r \frac{P}{K_{mP}}}{1 + \left(\frac{S}{K_{mS}} + \frac{P}{K_{mP}}\right) \cdot i_4}$	K_m and v_{max} change, but their ratio remains; S may not out compete I
Noncompetitive inhibition	I binds to E and ES; P-release only from ES $K_{I,3} = K_{I,4}, k_6 = 0$	$v = \dfrac{v_{max}S}{(K_m + S) \cdot i_3}$	$v = \dfrac{v_{max}^f \frac{S}{K_{mS}} - v_{max}^r \frac{P}{K_{mP}}}{\left(1 + \frac{S}{K_{mS}} + \frac{P}{K_{mP}}\right) \cdot i_4}$	K_m remains, v_{max} changes; S may not out compete I
Mixed inhibition	I binds to E and ES; P-release only from ES $K_{I,3} \neq K_{I,4}, k_6 = 0$	$v = \dfrac{v_{max}S}{K_m \cdot i_4 + S \cdot i_3}$		K_m and v_{max} change; $K_{I,3} > K_{I,4}$: competitive-noncompetitive inhibition; $K_{I,3} < K_{I,4}$: noncompetitive-competitive inhibition
Partial inhibition	I may bind to E and ES; P-release only from ES and ESI $K_{I,3} \neq K_{I,4}, k_6 \neq 0$	$v = \dfrac{v_{max}S \left(1 + \frac{k_6 \cdot I}{k_2 K_{I,3}}\right)}{K_m \cdot i_4 + S \cdot i_3}$		K_m and v_{max} change; if $k_6 > k_2$, activation instead of inhibition.

Abbreviations: $K_{I,3} = \frac{k_{-3}}{k_3}$, $K_{I,4} = \frac{k_{-4}}{k_4}$, $i_3 = 1 + \frac{I}{K_{I,3}}$, $i_4 = 1 + \frac{I}{K_{I,4}}$

enzyme class is unspecific and thus sensitive to substrate inhibition. Another consequence might be a drug-drug interaction where another drug is processed by the same enzyme as the main drug. This is why we talk about perpetrator and victim drugs—because of the influence of the pharmacokinetics over the other drug.

Cooperative enzymes

Regulatory enzymes often have more complicated kinetics and their sub-units display cooperative behavior. Cooperativity is the property whereby an enzyme can have a steep dependence on the substrate or inhibitor [18]. The positive[25] homotropic[26] cooperativity can be formulated on the following base for a dimeric enzyme E_2 [22]:

$$E_2 + S \xrightarrow[\text{slow}]{} E_2S$$
$$E_2S + S \xrightarrow[\text{fast}]{} E_2S_2$$

Because the second reaction is faster, one can assume that we have complete cooperativity. This means that the dimeric enzyme is either empty or full:

$$E_2 + 2S \longrightarrow E_2S_2$$

The binding constant:

$$K_B = \frac{E_2S_2}{E_2 \cdot S^2}$$

represents the equilibrium of concentration levels, and the fractional saturation of the enzyme is:

$$Y = \frac{2E_2S_2}{2E_{2,\text{total}}} = \frac{K_B S^2}{1 + K_B S^2}$$

However, this Hill[27] equation should be written with the parameter h (no integer) instead of n

$h < 1$ First ligand reduces affinity for the second ligand

$h = 1$ Quasi Michaelis-Menten

$h > 1$ First ligand increases affinity for the second ligand

It is incorrect to treat h as an estimate of the number of substrate-binding sites on the enzyme, though for some models it does provide a lower limit for this number [18]. Hemoglobin has 4 sub-units with a lower Hill coefficient, $h = 2.7$. Thus we reformulate our equation to:

$$Y = \frac{K_B S^h}{1 + K_B S^h}$$

We assume that the fractional enzyme saturation scales with the substrate turnover:

$$v = v_{\max} \frac{K_B S^h}{1 + K_B S^h}$$

Table 3. Calmodulin model. Different approaches to model cooperativity with and without simplifying assumptions. Source: [24]. Copyright © 2014, Elsevier B.V.

Model	Actual number of parameters (generated by framework)	Effective number of parameters (after assumptions)
Induced fit (generic)	15	15
Cooperative model (strong and weak sites)	15	5
Sequential	15	4
Allosteric	31	7
Macroscopic	4	4
Hill's equation	2	2

The kinetic dependence on the Hill coefficient is visualized in Figure 15.

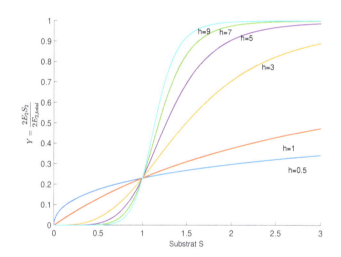

Figure 15. Hill kinetic. Factural saturation Y with $K_B = 0.3$ and different Hill coefficients.

The Hill equation is a pseudo-mechanistic equation which Hill himself saw as purely empirical. However, the use of the cooperativity index R_a of Taketa and Pogell is more empiric (experimentally meaningful) as it focusses on the 10-90% range of full activity. A mechanistic formulation is the Adair[28] equation, which is most often used for positive cooperativity in practice. Moreover, we have the allosteric model or symmetric model[29][30][31] with the assumption that the subunits of cooperative oligomeric proteins are independent of the ligand and all sub-units are in the same conformation at any time, while the se-

---

[25] Ligand facilitates binding of the next ligand. $h > 1$.

[26] Homotropic means that the allosteric modulator is the substrate. When the modulator is not the substrate it is called heterotropic.

[27] English physiologist Archibald Vivian Hill (1886—1977). Nobel Prize in medicine for his work on physical properties of muscles in 1922.

[28] British protein scientist Gilbert Smithson Adair (1896—1979).

[29] French biochemist Jacques Lucien Monod (1910—1976). Nobel Prize in Medicine for genetic control of enzyme and virus synthesis. Lac-operon-model.

[30] American biologist and biophysicist Jeffries Wyman (1901—1995).

[31] French neuroscientist Jean-Pierre Changeux (1936—today).

quential model[32] considers ligand-induced changes in a sub-unit with possible consequences to others. Thus, the sequential model assumes that sub-units show independent conformations in contrast to the allosteric MWC[33] model. Monomeric enzyme can show kinetic cooperativity if the different conformational states relax slowly. The section is based on the textbook found in [18]. Table 3 might give an impression of the model sizes for reflecting cooperativity.

---

[32] American biochemist Daniel Edward Koshland Jr. (1920—2007). Together with G. Némethy and D. Filmer in 1966.
[33] Author names: Monod-Wyman-Changeux.

2. Basics of Mathematics

We have a collection of mathematical methods to help us tackle biological questions. They are listed in Figure 16. We recommend this review for further reading [25].

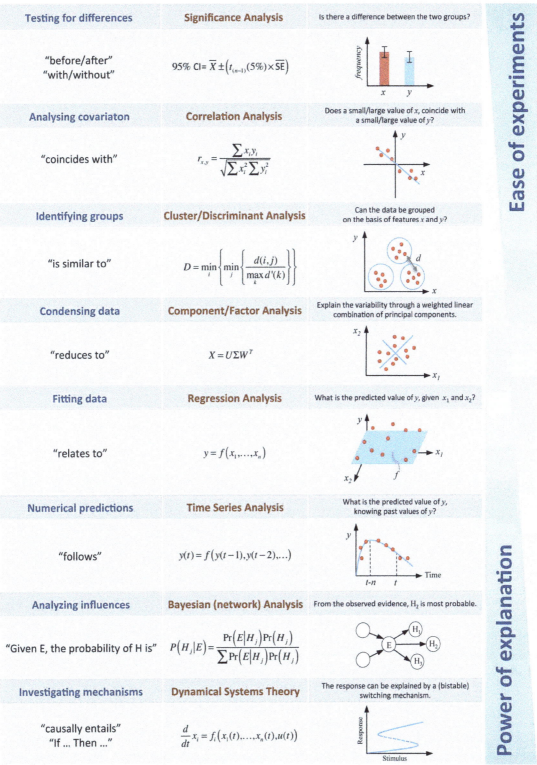

Figure 16. Mathematical analysis approaches. More information and source: [25]. Copyright © Frontiers Physiol, Licence: CC BY-SA 4.0.

References

[1] Sartre Jean-Paul. *La nausée*. 1938.

[2] Gregor Nickel, Markus Helmerich, Ralf Krömer, Katja Lengnink, and Martin Rathgeb. *Mathematik und Gesellschaft: Historische, philosophische und didaktische Perspektiven*. Springer-Verlag, 2018. https://doi.org/10.4000/essais.10180.

[3] George EP Box and Norman R Draper. *Empirical model-building and response surfaces*. John Wiley & Sons, 1987.

[4] Pieter Eykhoff. *System identification*. John Wiley & Sons, 1974.

[5] Frank Haußer and Yury Luchko. *Mathematische Modellierung mit MATLAB: Eine praxisorientierte Einführung*. Springer-Verlag, 2010.

[6] Werner Blum and Dominik Leiss. „Filling up"-the problem of independence-preserving teacher interventions in lessons with demanding modelling tasks. In *CERME 4–Proceedings of the Fourth Congress of the European Society for Research in Mathematics Education*, 2005.

[7] Tie Koide, Wyming Lee Pang, and Nitin S Baliga. The role of predictive modelling in rationally re-engineering biological systems. *Nature Reviews Microbiology*, 7(4):297, 2009. https://doi.org/10.1038/nrmicro2107.

[8] S Yu Jessica and Neda Bagheri. Multi-class and multi-scale models of complex biological phenomena. *Current Opinion in Biotechnology*, 39:167–173, 2016. https://doi.org/10.1016/j.copbio.2016.04.002.

[9] Stefano Schivo, Jetse Scholma, Paul E van der Vet, Marcel Karperien, Janine N Post, Jaco van de Pol, and Rom Langerak. Modelling with animo: Between fuzzy logic and differential equations. *BMC Systems Biology*, 10(1):56, 2016. https://doi.org/10.1186/s12918-016-0286-z.

[10] Paul Kirk, Thomas Thorne, and Michael PH Stumpf. Model selection in systems and synthetic biology. *Current Opinion in Biotechnology*, 24(4):767–774, 2013. https://doi.org/10.1016/j.copbio.2013.03.012.

[11] Daniel Silk, Paul DW Kirk, Chris P Barnes, Tina Toni, and Michael PH Stumpf. Model selection in systems biology depends on experimental design. *PLoS Computational Biology*, 10(6):e1003650, 2014. https://doi.org/10.1371/journal.pcbi.1003650.

[12] Sebastiaan H Mannaerts. Extensive quantities in thermodynamics. *European Journal of Physics*, 35(3):035017, 2014. https://doi.org/10.1088/0143-0807/35/3/035017.

[13] Roberto Marquardt, Juris Meija, Zoltán Mester, Marcy Towns, Ron Weir, Richard Davis, and Jürgen Stohner. Definition of the mole (iupac recommendation 2017). *Pure and Applied Chemistry*, 90(1):175–180, 2018. https://doi.org/10.1515/pac-2017-0106.

[14] F Peter Guengerich. Cytochrome p450 and chemical toxicology. *Chemical Research in Toxicology*, 21(1):70–83, 2007. https://doi.org/10.1021/tx700079z.

[15] Sameh Eid, Samo Turk, Andrea Volkamer, Friedrich Rippmann, and Simone Fulle. Kinmap: A web-based tool for interactive navigation through human kinome data. *BMC Bioinformatics*, 18(1):16, 2017. https://doi.org/10.1186/s12859-016-1433-7.

[16] Andrea Volkamer, Sameh Eid, Samo Turk, Friedrich Rippmann, and Simone Fulle. Identification and visualization of kinase-specific subpockets. *Journal of Chemical Information and Modeling*, 56(2):335–346, 2016. https://doi.org/10.1021/acs.jcim.5b00627.

[17] Robin E Ferner and Jeffrey K Aronson. Cato guldberg and peter waage, the history of the law of mass action, and its relevance to clinical pharmacology. *British Journal of Clinical Pharmacology*, 81(1):52–55, 2016. https://doi.org/10.1111/bcp.12721.

[18] Athel Cornish-Bowden. *Fundamentals of enzyme kinetics, 510*. Wiley-Blackwell Weinheim, Germany, 2014.

[19] Kenneth A Johnson and Roger S Goody. The original michaelis constant: Translation of the 1913 michaelis–menten paper. *Biochemistry*, 50(39):8264–8269, 2011. https://doi.org/10.1021/bi201284u.

[20] Boseung Choi, Grzegorz A Rempala, and Jae Kyoung Kim. Beyond the Michaelis-Menten equation: Accurate and efficient estimation of enzyme kinetic parameters. *Scientific Reports*, 7(1):17018, 2017. https://doi.org/10.1038/s41598-017-17072-z.

[21] Sascha Schäuble, Anne Kristin Stavrum, Pål Puntervoll, Stefan Schuster, and Ines Heiland. Effect of substrate competition in kinetic models of metabolic networks. *FEBS Letters*, 587(17):2818–2824, 2013. https://doi.org/10.1016/j.febslet.2013.06.025.

[22] Edda Klipp, Wolfram Liebermeister, Christoph Wierling, Axel Kowald, and Ralf Herwig. *Systems biology: A textbook*. John Wiley & Sons, 2016.

[23] Yuh Lin, Ping Lu, Cuyue Tang, Qin Mei, Grit Sandig, A David Rodrigues, Thomas H Rushmore, and Magang Shou. Substrate inhibition kinetics for cytochrome p450-catalyzed reactions. *Drug Metabolism and Disposition*, 29(4):368–374, 2001.

[24] Jacques Haiech, Yves Gendrault, Marie-Claude Kilhoffer, Raoul Ranjeva, Morgan Madec, and Christophe Lallement. A general framework improving teaching ligand binding to a macromolecule. *Biochimica et Biophysica Acta (BBA)-Molecular Cell Research*, 1843(10):2348–2355, 2014. https://doi.org/10.1016/j.bbamcr.2014.03.013.

[25] Olaf Wolkenhauer. Why model? *Frontiers in Physiology*, 5:21, 2014. https://doi.org/10.3389/fphys.2014.00021.

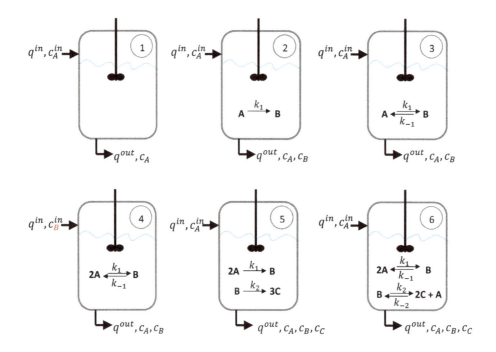

3. Exercises

◼ Balance equations

Set up balance equations for the volume and for the amount of substance (in mol) of every reaction partner for the systems depicted below. The reactions take place in an ideally mixed and closed reactor.

◼ CSTR

In the following, a Continuous Stirred Tank Reactor (CSTR) is investigated. A substance A with concentration c_A^{in} and temperature T^{in} is pumped into the reactor with the volume flow rate q^{in}. A 1st-order reaction takes place, during which Substance A is converted irreversibly into Substance B with a reaction rate constant k_1. To increase the conversion rate, heat flow $\dot{Q}_{heat}$ is fed into the CSTR. The change in reaction enthalpy is neglectable. The volume flow rate q^{out} (temperature T^{out}; concentrations c_A^{out}, c_B^{out}) is removed from the CSTR.

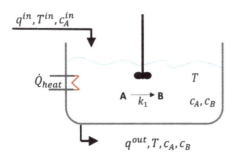

a) Set up the balance equations for the amounts of Substance A and Substance B.

For the questions below, the following simplifying assumptions shall apply: $q = q^{in} = q^{out}$ and $V =$ const.

b) Which simplified amount of substance balances can you achieve? Which variable are you balancing now?

c) Which concentrations c_A^*, c_B^* do you obtain in steady state?

◼ Balancing a CSTR

Cell-free medium (substrate S, influx concentration c_S^{in}, volume flow rate q) is added to a Continuous Stirred Tank Reactor (CSTR) in continuous cultivation mode. The reaction volume within the reactor (V_R) is kept constant. Micro-organisms are growing on the substrate within the CSTR. The specific growth rate μ of the micro-organisms is approximated with a Michaelis-Menten kinetic. The yield coefficient shall be Y_{XS}.

a) Set up mass balance equations for the concentrations of the biomass (X) and the substrate (S).

b) Derive the steady-state concentration values (c_S^*, c_X^*) as a function of the dilution rate $D = D^* = \dfrac{q}{V_R}$.

◼ Intracellular reaction

We will now focus on enzyme-catalyzed reactions in the interior of a bacterial cell with a constant cell volume V_C. Thereby a substrate S is converted into Product P:

$$S + E \underset{k_{-1}}{\overset{k_1}{\rightleftharpoons}} ES$$

$$ES \overset{k_2}{\longrightarrow} P + E$$

The reaction is regulated via a feedback repression of the enzyme synthesis. A simplified mechanism is assumed as follows:

$$C + D \xrightarrow{k_3} E + D \qquad \text{enzyme synthesis}$$

$$D + P \underset{k_{-4}}{\overset{k_4}{\rightleftharpoons}} DP \qquad \text{repression of enzyme synthesis}$$

$$E \xrightarrow{k_5} F \qquad \text{enzyme degradation}$$

Within the cell, ideal mixing is assumed. Reaction orders are according to the stoichiometric coefficients.

a) Set up balance equations for the concentrations of reaction partners E, ES, S, P, D, and DP in the form $\frac{d}{dt}(Vc)$. The bacterial cell is thereby considered as a closed system.

b) Show that the overall concentration of substance D is constant, i.e. $c_D + c_{DP} = \text{const} = c_{D0}$

c) Determine α, β and γ in:

$$\frac{dc_S}{dt} = -\alpha \cdot c_S \cdot c_E$$

$$\frac{dc_E}{dt} = \frac{\beta}{1 + \gamma \cdot c_P} - k_5 \cdot c_E$$

using the result of part b) and applying the following assumptions:

1. The concentration of the enzyme substrate complex c_{ES} is considered to be constant, i.e. $\frac{d}{dt} c_{ES} = 0$.

2. Reaction 4 is considered to be in equilibrium (quasi-steady state).

3. Concentration c_C is constant ($c_C = c_C^*$).

Remark: The results of points a) to c) are not needed for us to solve the following parts. For the remaining parts of this exercise, it is assumed that a constant substrate concentration c_S^* is achieved within the cell via regulation of the substrate transport:

$$c_S(t) = c_S^*$$

d) Which concentrations c_E^* and c_P^* do you get in steady state, if Product P is further converted with rate $r_P(t) = r_P^*$? This conversion results in the following equations:

$$\frac{dc_E}{dt} = \frac{\beta}{1 + \gamma \cdot c_P} - k_5 \cdot c_E \qquad \text{I}$$

$$\frac{dc_P}{dt} = \alpha \cdot c_S \cdot c_E - r_P \qquad \text{II}$$

e) Derive a single 2nd-order differential equation for the product concentration c_P from Equation I and II while applying the assumption $c_S = c_S^*$ and considering a time variable rate $r_P(t)$.

f) Linearize this differential equation around the steady state (c_P^*, r_P^*) with $\Delta c_P = c_P - c_P^*$ and $\Delta r_P = r_P - r_P^*$.

Mixed population

In the following we will focus on a mixed population in a CSTR in continuous cultivation. The reaction volume V is kept constant. The mixed population consists of two species of micro-organisms: prey (P) and predator (R). Prey grows with a specific growth rate μ and decreases proportionally with the number of predator and the number of prey. (The more prey is available, the more will be depredated). Ideal mixing and a constant growth rate μ (with $\mu \neq$ dilution rate $D = \frac{q}{V}$) is assumed.

a) Describe the occurring processes in the population.

b) Assign the individual terms within the following equations to the occurring population processes:

$$\frac{d}{dt} c_P = \mu c_P - k_1 c_P c_R - D c_P$$

$$\frac{d}{dt} c_R = k_2 c_R c_P - D c_R$$

c) Determine the steady state(s) and characterize it (them) qualitatively.

d) Linearize the system around the steady state(s).

e) Calculate the Eigenvalues for the cases:
   1. $D = 2\mu$
   2. $D = \frac{1}{2}\mu$

f) What can generally be stated on the stability of the steady states? Focus therefore on 3 cases ($\mu > D$, $\mu = D$ and $\mu < D$) for every steady state.

Deriving rate parameters from biological data

A cartoon model for gene regulation via an extracellular signaling molecule (also called a ligand) is given as follows:

L is the signaling molecule, R the membrane receptor, A an adaptor protein, and TF the transcription factor involved in the regulation.

From the cartoon model, one derives the biochemical

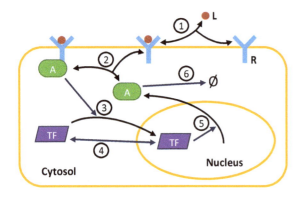

Cytosol

Nucleus

reaction network as follows:

$$L + R \underset{k_{-1}}{\overset{k_1}{\rightleftharpoons}} R_i \qquad \text{I}$$

$$R_i + A \underset{k_{-2}}{\overset{k_2}{\rightleftharpoons}} R_a \qquad \text{II}$$

$$R_a + TF \overset{k_3}{\longrightarrow} TFn \qquad \text{III}$$

$$TF \underset{k_{-4}}{\overset{k_4}{\rightleftharpoons}} TFn \qquad \text{IV}$$

$$TFn \overset{k_5}{\longrightarrow} TFn + A \qquad \text{V}$$

$$A \overset{k_6}{\longrightarrow} \qquad \text{VI}$$

The following information is known from biological measurements:

- Experiments are done with 2 mL of medium and $6 \cdot 10^5$ cells, using a ligand concentration of 1 nM.

- The volume of a cell is $1 \cdot 10^{-12} L$, with the nucleus taking $1/5$ of the total volume.

- The concentration of the transcription factor in whole cell extracts (TF + TFn) was measured as $0.1 \mu M$.

- The concentration of the transcription factor without stimulus in nuclear extracts was measured as 5nM.

- The concentration of the adaptor protein without stimulus in whole cell extracts is $1 \mu M$.

- The adaptor protein has a half-life time of 1 h.

- A cell has a total of 10000 membrane receptors.

- The associated constant K_a for the ligand-receptor binding is $0.01 \frac{1}{nM}$. Hint:

$$k_1 [L][R] = k_{-1}[Ri]$$

$$\Rightarrow \quad \frac{k_1}{k_{-1}} = \frac{[Ri]}{[L][R]} = K_a$$

- The associated constant for the receptor-adaptor protein binding is $0.01 \frac{1}{\mu M}$.

- The parameter k_3 has been measured directly as $k_3 = 0.01 \frac{1}{nM \cdot s}$.

The following assumptions, done by the modeler, are needed to determine all parameter values uniquely: $k_{-1} = 0.01 \frac{1}{s}$, $k_{-2} = 1 \cdot 10^{-3} \frac{1}{s}$, and $k_4 = 1 \cdot 10^{-5} \frac{1}{s}$.

a) Convert the unit molar M to molecules per cell for ligand and transcription factor states TF and TFn.

b) Calculate the degradation rate constant (k_6) from the half-life time $T_{1/2}$.

c) Construct an ODE model for the sub-network involving the reactions IV - VI, *i.e.*, the unstimulated case. Use, therefore, the law of mass action.

d) Use the biological measurements in the unstimulated case (steady-state condition) and the assumptions to determine all parameters in reactions IV - VI. Use units such that state variables are given as numbers of molecules per cell.

e) Determine the parameters for reactions I - III from the biological measurements and the assumptions, using the same units as before.

Notes

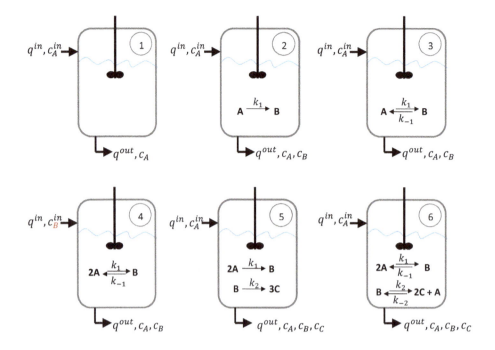

4. Solutions

Do not betray yourself!

Exercises

▪ Balance equations

Concentrations in the extraction pipe are assumed to correspond to the concentrations within the tank. Thus, we have omitted the specific indication with the indices *out*.

Reactor 1

$$\frac{dV}{dt} = q^{in} - q^{out}$$

$$\frac{dn_A}{dt} = n_A^{in} - n_A^{out}$$

$$= q^{in} c_A^{in} - q^{out} c_A$$

Reactor 2

$$\frac{dV}{dt} = q^{in} - q^{out}$$

$$\frac{dn_A}{dt} = q^{in} c_A^{in} - q^{out} c_A - k_1 c_A V$$

$$\frac{dn_B}{dt} = -q^{out} c_B + k_1 c_A V$$

Reactor 3

$$\frac{dV}{dt} = q^{in} - q^{out}$$

$$\frac{dn_A}{dt} = q^{in} c_A^{in} - q^{out} c_A - k_1 c_A V + k_{-1} c_B V$$

$$\frac{dn_B}{dt} = -q^{out} c_B + k_1 c_A V - k_{-1} c_B V$$

Reactor 4

$$\frac{dV}{dt} = q^{in} - q^{out}$$

$$\frac{dn_A}{dt} = -q^{out} c_A - 2k_1 c_A^2 V + 2k_{-1} c_B V$$

$$\frac{dn_B}{dt} = q^{in} c_B^{in} - q^{out} c_B + k_1 c_A^2 V - k_{-1} c_B V$$

Reactor 5

$$\frac{dV}{dt} = q^{in} - q^{out}$$

$$\frac{dn_A}{dt} = q^{in} c_A^{in} - q^{out} c_A - 2k_1 c_A^2 V$$

$$\frac{dn_B}{dt} = -q^{ab} c_B + k_1 c_A^2 V - k_2 c_B V$$

$$\frac{dn_C}{dt} = -q^{out} c_C + 3k_2 c_B V$$

Reactor 6

$$\frac{dV}{dt} = q^{in} - q^{out}$$

$$\frac{dn_A}{dt} = q^{in} c_A^{in} - q^{out} c_A - 2k_1 c_A^2 V + 2k_{-1} c_B V + k_2 c_B V$$
$$- k_{-2} c_C^2 c_A V$$

$$\frac{dn_B}{dt} = -q^{out} c_B + k_1 c_A^2 V - k_{-1} c_B V - k_2 c_B V + k_{-2} c_C^2 c_A V$$

$$\frac{dn_C}{dt} = -q^{out} c_C + 2k_2 c_B V - 2k_{-2} c_C^2 c_A V$$

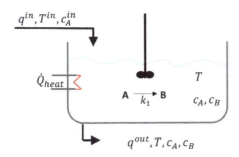

CSTR

Task a) Amount of substance balances:

$$\frac{dn_A}{dt} = q^{in}c_A^{in} - q^{out}c_A - k_1 c_A V$$

$$\frac{dn_B}{dt} = 0 - q^{out}c_B + k_1 c_A V$$

Task b) Applying the simplifying assumptions $q = q^{in} = q^{out}$ und $V = $ const.:

$$\frac{dn_A}{dt} = \frac{d}{dt}(c_A \cdot V) = \dot{V}c_A + V\dot{c}_A = 0 + V\dot{c}_A$$

Applying the product rule and assume that a CSTR does not change the volume in the tank during the process (derivative of constant V is zero):

$$\rightsquigarrow V \cdot \dot{c}_A = q c_A^{in} - q c_A - k_1 c_A V$$

$$\dot{c}_A = \frac{q}{V} \cdot (c_A^{in} - c_A) - k_1 \cdot c_A \qquad (4.1)$$

$$\dot{c}_B = -\frac{q}{V} \cdot c_B + k_1 c_A \qquad (4.2)$$

Task c) From the equation (4.1) with $\dot{c}_A = 0$:

$$\frac{q}{V}(c_A^{in} - c_A^*) - k_1 c_A^* \overset{!}{=} 0$$

$$\frac{q}{V}c_A^{in} - \frac{q}{V}c_A^*) - k_1 c_A^* = 0$$

$$\Rightarrow c_A^* = \frac{\frac{q}{V}(c_A^{in}}{k_1 + \frac{q}{V}}$$

$$c_A^* = \frac{q}{q + k_1 V}c_A^{in}$$

and for Substance B based on the equation (4.2) with $\dot{c}_B = 0$:

$$\dot{c}_B = -\frac{q}{V} \cdot c_B^* + k_1 c_A^* \overset{!}{=} 0$$

$$\frac{q}{V} \cdot c_B^* = k_1 c_A^*$$

$$c_B^* = \frac{k_1 V}{q}c_A^*$$

$$c_B^* = \frac{k_1 V}{q} \cdot \frac{q}{q + k_1 V}c_A^{in}$$

$$c_B^* = \frac{k_1 V c_A^{in}}{(q + k_1 V)}$$

Balancing a CSTR

Cell-free medium (substrate S, influx concentration c_S^{in}, volume flow rate q) is added to a Continuous Stirred Tank Reactor (CSTR) in continuous cultivation mode. The reaction volume within the reactor (V_R) is kept constant. Micro-organisms are growing on the substrate within the CSTR. The specific growth rate μ of the micro-organisms is approximated with a Michaelis-Menten kinetics. The yield coefficient shall be Y_{XS}.

Task a) Set up mass balance equations for the biomass (X) and the substrate (S) concentrations.

$$\underbrace{\frac{d(c_X V_R)}{dt}}_{\left[\frac{mol}{h}\right]} = \underbrace{-qc_X}_{\left[\frac{1}{h}\right]\left[\frac{mol}{l}\right]} + \underbrace{\mu c_X V_R}_{\left[\frac{1}{h}\right]\left[\frac{mol}{l}\right][l]} = -qc_X + \mu_{max}\frac{c_S}{K_S + c_S}c_X V_R$$

$$\frac{d(c_S V_R)}{dt} = q\left(c_S^{zu} - c_S\right) - \frac{1}{Y_{XS}}\mu c_X V_R$$

With $V_R = $ const. we get

$$\frac{dc_X}{dt} = c_X(\mu - \frac{q}{V}) = c_X(\mu - D) \qquad \text{I}$$

$$\frac{dc_S}{dt} = \frac{q}{V}(c_S^{zu} - c_S) - \frac{1}{Y_{XS}}\mu c_X = D(c_S^{zu} - c_S) - \frac{1}{Y_{XS}}\mu c_X \quad \text{II}$$

Task b) Derive the steady-state concentration values (c_S^*, c_X^*) as a function of the dilution rate $D = D^* = \frac{q}{V_R}$.

$$\text{Eq. I}: \frac{dc_X}{dt} = c_X(\mu - D) \overset{!}{=} 0$$

$$\underline{\text{Case 1:}} \Rightarrow \boxed{c_X^* = 0}$$

$$\rightarrow \text{in II}: D^*(c_S^* - c_S^*) \overset{!}{=} 0 \Rightarrow \boxed{c_S^* = c_S^{zu}}$$

$$\underline{\text{Case 2:}} \ \mu^* = \mu_{max}\frac{c_S^*}{c_S^* + K_S} = D^*$$

$$c_S^* \mu_{max} = D^* c_S^* + K_S D^*$$

$$\Rightarrow \boxed{c_S^* = \frac{K_S D^*}{\mu_{max} - D^*}} \qquad \text{III}$$

$$\rightarrow \text{in II}: D^*\left(c_S^{zu} - \frac{K_S D^*}{\mu_{max} - D}\right)$$

$$-\frac{1}{Y_{XS}}\mu_{max}\frac{\frac{K_S D^*}{\mu_{max} - D^*}}{K_S + \frac{K_S D^*}{\mu_{max} - D^*}}c_X^* \overset{!}{=} 0$$

$$\cancel{D^*}\left(c_S^{zu} - \frac{K_S D^*}{\mu_{max} - D}\right)$$

$$-\frac{\cancel{\mu_{max}}}{Y_{XS}}\frac{\cancel{K_S}\cancel{D^*}}{\cancel{K_S}\cancel{\mu_{max}}}c_X^* = 0$$

$$\Rightarrow \boxed{c_X^* = Y_{XS}\left(c_S^{zu} - \frac{K_S D^*}{\mu_{max} - D^*}\right)} \qquad \text{IV}$$

Conditions:
The substrate concentration cannot be negative ($C_S^* \geq 0$),

so that Equation III tells us that:

$$\Rightarrow \mu_{max} > D^* \qquad\qquad \text{V}$$

If $\mu_{max} < D^*$ we get a wash-out of the biomass, which is usually not desired in a CSTR.

Biomass should also be positive ($c_X^* > 0$):

$$\Rightarrow c_S^{zu} - \frac{K_S D^*}{\mu_{max} - D^*} > 0$$

with $\mu_{max} > D^*$:

$$c_S^{zu} \mu_{max} - c_S^{zu} D - K_S D^* > 0$$
$$\Rightarrow \mu_{max} > \frac{c_S^{zu} + K_S}{c_S^{zu}} D^* \qquad\qquad \text{VI}$$

whereby Equation V is contained in Equation VI, because:

$$\frac{c_S^{zu} + K_S}{c_S^{zu}} > 1. \qquad\qquad (4.3)$$

for practical considerations one uses $\mu_{max} > D$.

In summary:

Case 1: $\mu_{max} > \dfrac{C_S^{zu} + K_S}{C_S^{zu}} D^*$

$\Rightarrow$ Two steady states

Case 2: $\mu_{max} = \dfrac{C_S^{zu} + K_S}{C_S^{zu}} D^*$

$$\Rightarrow C_S^* = \frac{K_S D^*}{\frac{c_S^{zu} + K_S}{c_S^{zu}} D^* - D^*}$$
$$= \frac{K_S D^* C_S^{zu}}{(c_S^{zu} + K_S)D^* - D^* c_S^{zu}} = C_S^{zu}$$

$\Rightarrow c_X = 0$

$\Rightarrow$ Both steady states are equal.

Case 3: $\mu_{max} < \dfrac{C_S^{zu} + K_S}{C_S^{zu}} D^*$

$\Rightarrow$ It exists only the first steady state
$$c_X^* = 0; c_S* = c_S^{zu}$$

Intracellular reaction

We will now focus on enzyme-catalyzed reactions in the interior of a bacterial cell with a constant cell volume V_C. Thereby a substrate S is converted into Product P:

$$S + E \underset{k_{-1}}{\overset{k_1}{\rightleftharpoons}} ES$$

$$ES \xrightarrow{k_2} P + E$$

The reaction is regulated via a feedback repression of the enzyme synthesis. A simplified mechanism is assumed as follows:

$$C + D \xrightarrow{k_3} E + D \qquad \text{enzyme synthesis}$$

$$D + P \underset{k_{-4}}{\overset{k_4}{\rightleftharpoons}} DP \qquad \text{repression of enzyme synthesis}$$

$$E \xrightarrow{k_5} F \qquad \text{enzyme degradation}$$

Within the cell, ideal mixing is assumed. Reaction orders are according to the stoichiometric coefficients.

1st Part

Task a)   Set up balance equations for the concentrations of reaction partners E, ES, S, P, D, and DP in the form $\frac{d}{dt}(V \cdot c)$. The bacterial cell is thereby considered as a closed system.

$$\frac{d}{dt}(V_C \cdot c_E) = (-k_1 c_E c_S + k_{-1} c_{ES} + k_2 c_{ES} + k_3 c_D c_C - k_5 c_E)V_C$$
$$\frac{d}{dt}(V_C \cdot c_{ES}) = (k_1 c_E c_S - k_{-1} c_{ES} - k_2 c_{ES})V_C$$
$$\frac{d}{dt}(V_C \cdot c_S) = (-k_1 c_E c_S + k_{-1} c_{ES})V_C$$
$$\frac{d}{dt}(V_C \cdot c_P) = (k_2 c_{ES} - k_4 c_D c_P + k_{-4} c_{DP})V_C$$
$$\frac{d}{dt}(V_C \cdot c_D) = (-k_4 c_D c_P + k_{-4} c_{DP})V_C$$
$$\frac{d}{dt}(V_C \cdot c_{DP}) = (k_4 c_D c_P - k_{-4} c_{DP})V_C$$

with:

$$\frac{dn}{dt} = \frac{d(cV)}{dt} = V_C \frac{dc}{dt} + c \frac{dV_C}{dt} = V_C \frac{dc}{dt} \qquad \text{with } V_C = \text{const.}$$

Thus we get:

$$\frac{d}{dt} c_E = -k_1 c_E c_S + k_{-1} c_{ES} + k_2 c_{ES} + k_3 c_D c_C - k_5 c_E \qquad \text{I}$$
$$\frac{d}{dt} c_{ES} = k_1 c_E c_S - k_{-1} c_{ES} - k_2 c_{ES} \qquad \text{II}$$
$$\frac{d}{dt} c_S = -k_1 c_E c_S + k_{-1} c_{ES} \qquad \text{III}$$
$$\frac{d}{dt} c_P = k_2 c_{ES} - k_4 c_D c_P + k_{-4} c_{DP} \qquad \text{IV}$$
$$\frac{d}{dt} c_D = -k_4 c_D c_P + k_{-4} c_{DP} \qquad \text{V}$$
$$\frac{d}{dt} c_{DP} = k_4 c_D c_P - k_{-4} c_{DP} \qquad \text{VI}$$

which works only if the volume remains constant.

Task b)   Show that the overall concentration of Substance D is constant, i.e. $c_D + c_{DP} = \text{const} = c_{D0}$

If you look at the balances of c_D and c_{DP}, you see that the sum is zero:

$$\frac{d}{dt} c_D + \frac{d}{dt} c_{DP} = -k_4 c_D c_P + k_{-4} c_{DP} + k_4 c_D c_P - k_{-4} c_{DP} = 0.$$

By integrating this equation, one obtains:

$$\boxed{c_D + c_{DP} = \text{const.}}$$

Task c) Determine α, β, and γ in:

$$\frac{dc_S}{dt} = -\alpha \cdot c_S \cdot c_E \qquad\qquad \text{VII}$$

$$\frac{dc_E}{dt} = \frac{\beta}{1 + \gamma \cdot c_P} - k_5 \cdot c_E \qquad\qquad \text{VIII}$$

using the result of part b) and applying the following assumptions:

1. The concentration of the enzyme substrate complex c_{ES} is considered to be constant, i.e. $\frac{d}{dt}c_{ES} = 0$:

$$\frac{d}{dt}c_{ES} = k_1 c_E \cdot c_S - k_{-1}c_{ES} - k_2 c_{ES} \overset{!}{=} 0$$

$$c_{ES} = \frac{k_1}{k_{-1} + k_2}c_E c_S \qquad\qquad \text{IX}$$

in Equation III:

$$\frac{dS}{dt} = -k_1 c_E c_S + \frac{k_{-1}k_1}{k_{-1} + k_2}c_E c_S$$

$$= \frac{-k_1 k_{-1} - k_1 k_2 + k_{-1}k_1}{k_{-1} + k_2}c_E c_S$$

$$\Leftrightarrow \frac{dS}{dt} = -\frac{k_1 k_2}{k_{-1} + k_2}c_E c_S$$

$$\Rightarrow \boxed{\alpha = \frac{k_1 k_2}{k_{-1} + k_2}}$$

2. Reaction IV is considered to be in equilibrium (quasi-steady state).
   We already have α and still need γ and β. These 2 parameters can be found in Equation VIII, because this equation contains the derivative of c_E. We might tinker something together that looks similar. Let's use Reaction I:

$$\frac{dc_E}{dt} = -k_1 c_E c_S + (k_{-1} + k_2)c_{ES} + k_3 c_D c_C - k_5 c_E$$

Now, we have the problem that we have the terms c_E and c_S, and c_{ES} (which we do not want). If we put Equation IX in we get:

$$\frac{dc_E}{dt} = -k_1 c_E c_S + \frac{k_1 \cancel{(k_{-1} + k_2)}}{\cancel{k_{-1} + k_2}}c_E c_S$$
$$+ k_3 c_D c_C - k_5 c_E$$

$$\frac{dc_E}{dt} = k_3 c_D c_C - k_5 c_E$$

Well, that is some progress. Now, we take care of c_D, c_C and therefore we have another hint in the task description. Let's use the equilibrium equation (IV):

$$\frac{d}{dt}c_P = k_2 c_{ES} - k_4 c_D c_P + k_{-4}c_{DP} \overset{!}{=} 0$$

$$k_4 c_D c_P = k_{-4}c_{DP} + k_2 c_{ES}$$

We got another hint from task b) where we define a total concentration of enzyme-regulating protein c_{D0} with $c_D + c_{DP} = c_{D0} = \text{const.}$ Using this gives:

$$k_4 c_D c_P = k_{-4}(c_{D0} - c_D) + k_2 c_{ES}$$

$$(k_4 c_P + k_{-4})c_D = k_{-4}c_{D0} + k_2 c_{ES}$$

$$c_D = \frac{k_{-4}c_{D0} + k_2 c_{ES}}{(k_4 c_P + k_{-4})}$$

$$= \frac{c_{D0} + \frac{k_2}{k_{-4}}c_{ES}}{(\frac{k_4}{k_{-4}}c_P + 1)}$$

$$c_{DP} = \frac{k_4}{k_{-4}}c_D c_P$$

3. Concentration c_C is constant ($c_C = c_C^*$).

Finally, we obtain the parameters:

$$\alpha = \frac{k_1 k_2}{k_{-1} + k_2}$$

$$\gamma = \frac{k_4}{k_{-4}}$$

$$\beta = k_3 c_C^* c_{D0}$$

2nd Part

We assume now that we give as much substrate as the cells consume so that:

$$c_S(t) = c_S^*.$$

Task d) Which concentrations c_E^* and c_P^* do you get in steady state, if Product P is further converted with rate $r_P(t) = r_P^*$? This conversion results in the following equations:

$$\frac{dc_E}{dt} = \frac{\beta}{1 + \gamma \cdot c_P} - k_5 \cdot c_E \qquad\qquad \text{I}$$

$$\frac{dc_P}{dt} = \alpha \cdot c_S \cdot c_E - r_P \qquad\qquad \text{II}$$

In the stationary case:

$$\frac{dc_P}{dt} = \alpha \cdot c_S \cdot c_E - r_P \overset{!}{=} 0 \quad \Rightarrow \quad \boxed{c_E^* = \frac{r_P^*}{\alpha \cdot c_S^*}}$$

$$\frac{dc_E}{dt} = \frac{\beta}{1 + \gamma \cdot c_P} - k_5 \cdot c_E \overset{!}{=} 0$$

$$\frac{\beta}{1 + \gamma \cdot c_P^*} = k_5 \cdot c_E^* = k_5 \cdot \frac{r_P^*}{\alpha \cdot c_S^*}$$

$$1 + \gamma \cdot c_P^* = \frac{\alpha \beta c_S^*}{k_5 \cdot r_P^*}$$

$$\Rightarrow c_P^* = \frac{\alpha \beta c_S^*}{\gamma k_5 \cdot r_P^*} - \frac{1}{\gamma} = \frac{\alpha \beta c_S^* - k_5 \cdot r_P^*}{\gamma k_5 \cdot r_P^*}$$

Task e)   Derive a single 2nd-order differential equation for the product concentration c_P from Equation I and II while applying the assumption $c_S = c_S^*$ and considering a time variable rate $r_P(t)$. Assume $c_S(t) = c_S^*$.

From the system equations I and II we get:

$$\frac{d}{dt} c_E + k_5 c_E = \frac{\beta}{1 + \gamma c_P}$$

$$c_E = \frac{1}{\alpha c_S^*} \left(\frac{d}{dt} c_P + r_P \right)$$

We use this to obtain:

$$\frac{1}{\alpha c_S^*} \left(\frac{d^2}{dt^2} c_P + \frac{d}{dt} r_P \right) + \frac{k_5}{\alpha c_S^*} \left(\frac{d}{dt} c_P + r_P \right) = \frac{\beta}{1 + \gamma c_P}$$

or

$$\boxed{\frac{d^2}{dt^2} c_P + k_5 \frac{d}{dt} c_P - \frac{\alpha c_S^* \beta}{1 + \gamma c_P} = -\frac{d}{dt} r_P - k_5 r_P}.$$

Task f)   Linearize this differential equation around the steady state (c_P^*, r_P^*) with $\Delta c_P = c_P - c_P^*$ and $\Delta r_P = r_P - r_P^*$. Assume $c_S(t) = c_S^*$.

The linearization of $f = \frac{1}{1 + \gamma c_P}$ delivers:

$$f(c_P^*) + \frac{\partial f}{\partial c_P}\Big|_{c_P^*} \Delta c_P = \frac{1}{1 + \gamma c_P^*} + \frac{\partial f}{\partial c_P}\Big|_{c_P^*} \Delta c_P$$

$$= \frac{1}{1 + \gamma c_P^*} - \frac{1}{(1 + \gamma c_P^*)^2} \cdot \gamma \cdot \delta c_P \ldots,$$

with which we get:

$$\frac{d^2}{dt^2} \Delta c_P + k_5 \frac{d}{dt} \Delta c_P - k_5 r_P^* + \frac{(k_5 \cdot r_P^*)^2 \gamma}{\alpha \beta c_S^*} \Delta c_P$$

$$= -\frac{d}{dt} \Delta r_P - k_5 (r_P^* + \Delta r_P),$$

and finally:

$$\boxed{\frac{d^2}{dt^2} \Delta c_P + k_5 \frac{d}{dt} \Delta c_P + \frac{(k_5 \cdot r_P^*)^2 \gamma}{\alpha \beta c_S^*} \Delta c_P = -\frac{d}{dt} \Delta r_P - k_5 \Delta r_P}.$$

Mixed population

In the following we will focus on a mixed population in a CSTR in continuous cultivation. The reaction volume V is kept constant. The mixed population consists of two species of micro-organisms: prey (P) and predator (R). Prey grows with a specific growth rate μ and decreases proportionally with the number of predator and the number of prey. (The more prey is available, the more will be depredated). Ideal mixing and a constant growth rate μ (with $\mu \neq$ dilution rate $D = \frac{q}{V}$) is assumed.

Task a)   Describe the occurring processes in the population.

Birth, prey-predator interaction, extraction

Task b)   Assign the individual terms within the following equations to the occurring population processes:

$$\text{prey concentration:} \quad \frac{d}{dt} c_P = \overset{\text{birth}}{\mu c_P} \quad \overset{\text{predation}}{-k_1 c_R c_P} \quad \overset{\text{extraction}}{-D c_P}$$

$$\text{predator concentration:} \quad \frac{d}{dt} c_R = \underset{\text{growth on prey}}{k_2 c_P c_R} \quad \underset{\text{extraction}}{-D c_R}$$

Task c)   Determine the steady state(s) and characterize it (them) qualitatively.

$$0 = (\mu - k_1 c_R^* - D) c_P^* \qquad \text{I}$$
$$0 = (k_2 c_B^* - D) c_R^* \qquad \text{II}$$

Steady State 1: $\rightarrow$ $\boxed{c_{P1}^* = 0, c_{R1}^* = 0}$

Steady State 2: $\rightarrow$ $\boxed{c_{P2}^* = \frac{D}{k_2}, c_{R1}^* = \frac{\mu - D}{k_1}}$

Steady State 1 basically says that neither prey nor predators are present because they are washed out.
Steady State 2 exists only if $c_{R2}^* \geq 0$ is. Therefore, the growth rate has to compensate at least the dilution rate $\mu \geq D$. If $\mu \leq D$, we have Steady state 1.

Task d)   Linearize the system around the steady state(s).

From Equation I:

$$\dot{c}_P = \mu c_B - k_1 c_P c_R - D c_P$$

we get:

$$\underbrace{\dot{c}_P^*}_{0} + \Delta \dot{c}_P = \underbrace{f(c_B^*, c_R^2)}_{0(ss)} + (\mu - D) \cdot \Delta c_P - k_1 c_P^* \cdot \Delta c_R - k_1 c_R^* \cdot \Delta c_P$$

$$\Delta \dot{c}_P = (\mu - k_1 c_R^* - D) \cdot \Delta c_P - k_1 c_P^* \cdot \Delta c_R$$

with $c_P = c_P^* + \Delta c_B$ and $c_R = c_R^* + \Delta c_R$.

From Equation II:

$$\dot{c}_R = k_2 c_P c_R - D c_R$$

we get:

$$\Delta \dot{c}_R = -D \cdot \Delta c_R + k_2 c_R^* \cdot \Delta c_P + k_2 c_P^* \cdot \Delta c_R$$
$$\Delta \dot{c}_R = k_2 c_R^* \cdot \Delta c_P + (k_2 c_P^* - D) \cdot \Delta c_R$$

Steady State 1 (0,0):

$$\Delta \dot{c}_P = (\mu - D) \cdot \Delta c_P$$
$$\Delta \dot{c}_R = -D \cdot \Delta c_R$$
$$\begin{pmatrix} \Delta \dot{c}_P \\ \Delta \dot{c}_R \end{pmatrix} = \begin{pmatrix} \mu - D & 0 \\ 0 & -D \end{pmatrix} \begin{pmatrix} \Delta c_P \\ \Delta c_R \end{pmatrix}$$

Steady State 2 $\left(\frac{D}{k_2}, \frac{\mu - D}{k_1} \right)$:

$$\Delta \dot{c}_P = (\mu - \mu + D - D) \cdot \Delta c_P - \frac{k_1}{k_2} D \Delta c_R = -\frac{k_1}{k_2} D \Delta c_R$$

$$\Delta \dot{c}_R = \frac{k_2}{k_1} (\mu - D) \Delta c_P + (D - D) \cdot \Delta c_R = \frac{k_2}{k_1} (\mu - D) \cdot \Delta c_P$$

$$\begin{pmatrix} \Delta \dot{c}_P \\ \Delta \dot{c}_R \end{pmatrix} = \begin{pmatrix} \mu 0 & -\frac{k_1}{k_2} D \\ \frac{k_2}{k_1} (\mu - D) & 0 \end{pmatrix} \begin{pmatrix} \Delta c_P \\ \Delta c_R \end{pmatrix}$$

Task e) Calculate the Eigenvalues for the cases:
1. $D = 2\mu$
2. $D = \frac{1}{2} \mu$

Steady State 1 (0,0):

$$|sI - A| = 0$$
$$[s - (\mu - D)](s + D) = 0 \quad \Rightarrow \boxed{s_1 = \mu - D; s_2 = -D}$$

1.) $D = 2\mu$

$$s_1 = -\mu$$
$$s_2 = -2\mu$$

$\Rightarrow$ stable

2.) $D = \frac{1}{2} \mu$

$$s_1 = \frac{1}{2} \mu$$
$$s_2 = -\frac{1}{2} \mu$$

$\Rightarrow$ unstable saddle

Steady State 2 $\left(\frac{D}{k_2}, \frac{\mu - D}{k_1} \right)$:

$$|sI - A| = | \begin{bmatrix} s & 0 \\ 0 & s \end{bmatrix} - \begin{bmatrix} 0 & -\frac{k_1}{k_2} D \\ k_2 \frac{\mu - D}{k_1} & 0 \end{bmatrix} | \overset{!}{=} 0$$

$$s^2 + D(\mu - D) = 0$$

$$\boxed{s_{1,2} = \pm \sqrt{-D(\mu - D)}}$$

1.) $D = 2\mu$ Because the growth rate is just the half of the dilution rate we will have a wash-out. See task c).

2.) $D = \frac{1}{2} \mu$

$$s_1 = \frac{1}{2} \mu$$
$$s_2 = -\frac{1}{2} \mu$$

$\Rightarrow$ unstable saddle
Steady State 2 $\left(\frac{D}{k_2}, \frac{\mu - D}{k_1} \right)$:

$$\Rightarrow s_{1,2} = \pm \sqrt{-\frac{1}{2} \mu \cdot \frac{1}{2} \mu} = \pm i \cdot \frac{1}{2} \cdot \mu$$

$\Rightarrow$ metastable circle (limit circle).

Task f) What can generally be stated on the stability of the steady states? Focus therefore on 3 cases ($\mu > D$, $\mu = D$ and $\mu < D$) for every steady state.

Steady State 1	$\mu > D$	$\mu = D$	$\mu < D$
unstable	X		
stable			X
metastable		X	
Steady State 2	$\mu > D$	$\mu = D$	$\mu < D$
unstable		X	not exists
stable			not exists
metastable	X		not exists

■ **Deriving rate parameters from biological data**
A cartoon model for gene regulation via an extracellular signaling molecule (also called a ligand) is given as follows:

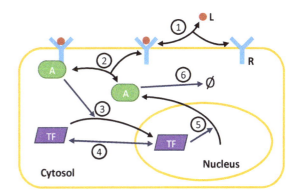

L is the signaling molecule, R the membrane receptor, A an adaptor protein, and TF the transcription factor involved in the regulation.

From the cartoon model, one derives the biochemical

reaction network:

$$L + R \underset{k_{-1}}{\overset{k_1}{\rightleftharpoons}} R_i \qquad \text{I}$$

$$R_i + A \underset{k_{-2}}{\overset{k_2}{\rightleftharpoons}} R_a \qquad \text{II}$$

$$R_a + TF \overset{k_3}{\longrightarrow} TFn \qquad \text{III}$$

$$TF \underset{k_{-4}}{\overset{k_4}{\rightleftharpoons}} TFn \qquad \text{IV}$$

$$TFn \overset{k_5}{\longrightarrow} TFn + A \qquad \text{V}$$

$$A \overset{k_6}{\longrightarrow} \qquad \text{VI}$$

The following information is known from biological measurements:

- Experiments are done with 2 mL of medium and $6 \cdot 10^5$ cells, using a ligand concentration of 1 nM.

- The volume of a cell is $1 \cdot 10^{-12}L$, with the nucleus taking $1/5$ of the total volume.

- The concentration of the transcription factor in whole cell extracts (TF + TFn) was measured as $0.1 \mu M$.

- The concentration of the transcription factor without stimulus in nuclear extracts was measured as 5nM.

- The concentration of the adaptor protein without stimulus in whole cell extracts is $1 \mu M$.

- The adaptor protein has a half-life time of 1 h.

- A cell has a total of 10000 membrane receptors.

- The associated constant K_a for the ligand-receptor binding is $0.01 \frac{1}{nM}$. Hint:

$$k_1 [L][R] = k_{-1}[Ri]$$
$$\Rightarrow \quad \frac{k_1}{k_{-1}} = \frac{[Ri]}{[L][R]} = K_a$$

- The associated constant for the receptor-adaptor protein binding is $0.01 \frac{1}{\mu M}$.

- The parameter k_3 has been measured directly as: $k_3 = 0.01 \frac{1}{nM \cdot s}$

The following assumptions, done by the modeler, are needed to determine all parameter values uniquely: $k_{-1} = 0.01 \frac{1}{s}$, $k_{-2} = 1 \cdot 10^{-3} \frac{1}{s}$, $k_4 = 1 \cdot 10^{-5} \frac{1}{s}$

Task a) Convert the unit molar M to molecules per cell for ligand and transcription factor TF and TFn.
Ligand:

$$[L] = 1nM = 10^{-9} \frac{mol}{L} = 6 \cdot 10^{14} \frac{molecules}{L}$$

$$\Rightarrow n_L = c \cdot V = 6 \cdot 10^{14} \frac{molecules}{L} \cdot 2 \cdot 10^{-3} L$$

$$= 1.2 \cdot 10^{12} \text{ molecules in 2 mL}$$

$$\Rightarrow \frac{1.2 \cdot 10^{12} \text{molecules}}{6 \cdot 10^5 \text{cells}} = \boxed{2 \cdot 10^6 \frac{molecules}{cell}}$$

Transcription factor:

$$[TFn] = 5nM = 5 \frac{nmol}{L} = 5 \cdot 10^{-9} \frac{mol}{L}$$

$$\Rightarrow 5 \cdot 10^{-9} \frac{mol}{L} \cdot N_A \cdot V$$

$$= 5 \cdot 10^{-9} \frac{mol}{L} \cdot 6.022 \cdot 10^{23} \frac{molecules}{mol} \cdot 0.2 \cdot 10^{-12} L$$

$$= \boxed{602 \frac{molecules}{cell}}$$

Task b) Calculate the degradation rate constant k_6 from the half-life time $T_{1/2}$:

$$\dot{A} = -k_6 \cdot A$$

$$\Rightarrow A(t) = A_0 \cdot e^{-k_6 \cdot t}$$

$$T_{1,2}: \quad A(t) \overset{!}{=} A_0/2$$

$$\Rightarrow \frac{A_0}{2} = A_0 \cdot e^{-k_6 \cdot T_{1/2}}$$

$$\frac{1}{2} = e^{-k_6 \cdot T_{1/2}}$$

$$ln\left(\frac{1}{2}\right) = -k_6 \cdot T_{1/2}$$

$$\underbrace{ln(1)}_{0} - ln(2) = -k_6 \cdot T_{1/2}$$

$$\Rightarrow k_6 = \frac{ln(2)}{T_{1/2}}$$

Task c) Construct an ODE model for the sub-network involving the reactions IV - VI, i.e. the unstimulated case. Use, therefore, the law of mass action.

$$\dot{c}_L = -k_1 c_L c_R + k_{-1} c_{Ri} \qquad \text{I}$$
$$\dot{c}_R = -k_1 c_L c_R + k_{-1} c_{Ri} \qquad \text{II}$$
$$\dot{c}_{Ri} = +k_1 c_L c_R - k_{-1} c_{Ri} - k_2 c_{Ri} c_A + k_{-2} c_{Ra} \qquad \text{III}$$
$$\dot{c}_{Ra} = +k_2 c_{Ri} c_A - k_{-2} c_{Ra} - k_3 c_{Ra} c_{TF} \qquad \text{IV}$$
$$\dot{c}_A = -k_2 c_{Ri} c_A + k_{-2} c_{Ra} + k_5 c_{TFn} - k_6 c_A \qquad \text{V}$$
$$\dot{c}_{TF} = -k_3 c_{Ra} c_{TF} - k_4 c_{TF} + k_{-4} c_{TFn} \qquad \text{VI}$$
$$\dot{c}_{TFn} = +k_3 c_{Ra} c_{TF} + k_4 c_{TF} - k_{-4} c_{TFn} \qquad \text{VII}$$

Task d) Use the biological measurements in the unstimulated case (steady-state condition) and the assumptions to determine all parameters in reactions IV - VI. Use units such that state variables are given as numbers of molecules per cell.

$$I = II: \qquad -k_1 \cdot 0 \cdot c_R^* + k_{-1} c_{Ri}^* = 0$$

$$\Rightarrow \boxed{c_{Ri}^* = 0}$$

$$III: \qquad k_1 \cdot 0 \cdot c_R^* - k_{-1} \cdot 0 - k_2 \cdot 0 + k_{-2} \cdot c_{Ra}^* = 0$$

$$\Rightarrow \boxed{c_{Ra}^* = 0}$$

$$IV: \qquad k_2 \cdot 0 \cdot c_A - k_{-2} \cdot 0 - k_3 \cdot 0 \cdot c_{TF} = 0$$

$$V.2: \qquad \dot{c}_A = -k_2 c_{Ri} c_A + k_{-2} c_{Ra} + k_5 c_{TFn} - k_6 c_A = 0$$

$$VI.2: \qquad \dot{c}_{TF} = -k_3 c_{Ra} c_{TF} - k_4 c_{TF} + k_{-4} c_{TFn} = 0$$

$$VII.2: \qquad \dot{c}_{TFn} = +k_3 c_{Ra} c_{TF} + k_4 c_{TF} - k_{-4} c_{TFn} = 0$$

Volume:

given: $\qquad V_{Cell} = 10^{-12} L$

$$\Rightarrow V_{nucl} = \frac{1}{5} \cdot 10^{-12} L = 0.2 \cdot 10^{-12} L$$

$$\Rightarrow V_{cyto} = \frac{4}{5} \cdot 10^{-12} L = 0.8 \cdot 10^{-12} L$$

Concentrations:

given: $c_{TF} + c_{TFn} = 0.1 \mu M$

$$\underbrace{\Rightarrow}_{c_{TF} + c_{TFn}} = 0.1 \cdot 10^{-6} \frac{mol}{L} \cdot N_A \cdot V_{cell}$$

$$= 0.1 \cdot 10^{-6} \frac{mol}{L} \cdot 6.022 \cdot 10^{23} \frac{molecules}{cell} \cdot 10^{-12} L$$

$$= 60220 \frac{molecules}{cell}$$

given: $c_{TFn} = 5nM$

$$\Rightarrow = 5 \cdot 10^{-9} \frac{mol}{L} \cdot N_A \cdot V_{nucl}$$

$$= 5 \cdot 10^{-9} \frac{mol}{L} \cdot 6.022 \cdot 10^{23} \frac{molecules}{cell} \cdot 0.2 \cdot 10^{-12} L$$

$$= 602 \frac{molecules}{cell}$$

$$\Rightarrow c_{TF} = 60220 - 602 = 59618 \frac{molecules}{cell}$$

given: $c_A = 1\mu M$

$$\Rightarrow c_A = 1 \cdot 10^{-6} \frac{mol}{L} \cdot N_A \cdot V_{cell} = 602200 \frac{molecules}{cell}$$

given: $k_4 = 10^{-5} \frac{1}{s}$

$$VI.2 :\Rightarrow \quad k_{-4} = k_4 \cdot \frac{c_{TF}}{c_{TFn}}$$

$$= 10^{-5} \frac{1}{s} \cdot \frac{59618}{602} = 9.9 \cdot 10^{-4} \frac{1}{s}$$

given: $T_{1/2}(A) = 1h = 3600s$

$$\Rightarrow \quad k_6 = \frac{ln(2)}{T_{1/2}(A)} = \frac{ln(2)}{3600s} = 1.9 \cdot 10^{-4} \frac{1}{s}$$

given: $k_4 = 10^{-5} \frac{1}{s}$

$$V.2 :\Rightarrow \quad k_5 = k_6 \cdot \frac{c_A}{c_{TFn}}$$

$$= 1.9 \cdot 10^{-4} \frac{1}{s} \cdot \frac{602000}{602} = 0.19 \frac{1}{s}$$

Task e) Determine the parameters for reactions I - III from the biological measurements and the assumptions, using the same units as before.

given: $k_a(L - R) = 0.01 \frac{1}{nM}$, $k_{-1} = 0.01 \frac{1}{s}$

$$\Rightarrow K_a = \frac{k_1}{k_{-1}} = 0.01 \frac{1}{nM} \quad \left(Lig.: 1nM \hat{=} 2 \cdot 10^6 \frac{molec.}{cell}\right)$$

$$K_a = \frac{k_1}{k_{-1}} = 5 \cdot 10^{-9} \frac{cell}{molecules}$$

$$k_1 = k_{-1} \cdot 5 \cdot 10^{-9} \frac{cell}{molecules}$$

$$= 0.01 \frac{1}{s} \cdot 5 \cdot 10^{-9} \frac{cell}{molecules}$$

$$= 5 \cdot 10^{-11} \frac{cell}{molecules}$$

given: $K_a(R_i - A) = 0.1 \frac{1}{\mu M}$ considerd only in the cytoplasma V_{cyto}. $k_{-2} = 10^{-3} \frac{1}{s}$.

$$\Rightarrow K_a = \frac{k_2}{k_{-2}} = 0.1 \frac{L}{10^{-6} mol} \cdot \frac{1}{N_A \cdot V_{cyto}}$$

$$= 0.1 \cdot 10^6 \frac{1}{6.022 \cdot 10^{23} \cdot 0.8 \cdot 10^{-12}} \frac{L}{mol} \cdot \frac{mol}{molecules} \cdot \frac{cell}{L}$$

$$= 2.1 \cdot 10^{-7} \frac{cell}{molecules}$$

$$\Rightarrow k_2 = K_a \cdot k_{-2} = 2.1 \cdot 10^{-7} \frac{cell}{molecules} \cdot 10^{-3} \frac{1}{s}$$

$$= 2.1 \cdot 10^{-10} \frac{cell}{s \cdot molecules}$$

given: $k_3 = 0.01 \frac{1}{nM \cdot s}$ considered only in the cytoplasma.

$$\Rightarrow k_3 = 0.01 \frac{L}{10^{-9} mol \cdot s} \cdot \frac{mol}{6.022 \cdot 10^{23} molecules} \cdot \frac{cell}{0.8 \cdot 10^{-12} L}$$

$$= 2.1 \cdot 10^{-5} \frac{cell}{molecules \cdot s}$$

Parameter	Value	Unit
K_a	$2.1 \cdot 10^{-7}$	cell/molecules
k_1	$5 \cdot 10^{-11}$	cell/s·molec
k_{-1}	0.01	$1/s$
k_2	$2.1 \cdot 10^{-10}$	cell/s·molec
k_{-2}	10^{-3}	$1/s$
k_3	$2.1 \cdot 10^{-5}$	cell/s·molec
k_4	10^{-5}	$1/s$
k_{-4}	10^{-3}	$1/s$
k_5	0.19	$1/s$
k_6	$1.9 \cdot 10^{-4}$	$1/s$

Notes

Notes

This book need not end here...

Share

All our books — including the one you have just read — are free to access online so that students, researchers and members of the public who can't afford a printed edition will have access to the same ideas. This title will be accessed online by hundreds of readers each month across the globe: why not share the link so that someone you know is one of them?

This book and additional content is available at:

https://doi.org/10.11647/OBP.291

Donate

Open Book Publishers is an award-winning, scholar-led, not-for-profit press making knowledge freely available one book at a time. We don't charge authors to publish with us: instead, our work is supported by our library members and by donations from people who believe that research shouldn't be locked behind paywalls.

Why not join them in freeing knowledge by supporting us: https://www.openbookpublishers.com/support-us

Follow @OpenBookPublish

Read more at the Open Book Publishers BLOG

You may also be interested in:

Making up Numbers
A History of Invention in Mathematics
Ekkehard Kopp

https://doi.org/10.11647/OBP.0236

 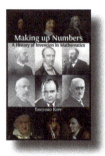

Animals and Medicine
The Contribution of Animal Experiments to the Control of Disease
Jack Bottin

https://doi.org/10.11647/OBP.0055

B C, Before Computers
On Information Technology from Writing to the Age of Digital Data
Stephen Robertson

https://doi.org/10.11647OBP.0225

www.ingramcontent.com/pod-product-compliance
Lightning Source LLC
LaVergne TN
LVHW061924050326
832903LV00038B/4831